PRAISE FOR
THE TAO OF LEADERSHIP

In *The Tao of Leadership*, Jack Myers offers a wide variety of concrete, implementable strategies that leaders can use to adapt their business practices and ensure that they are attuned with the needs of their workforce and the shifts in their markets in the age of AI and increased economic automation. Brimming with anecdotes and brief case studies, this book can provide a valuable desk reference in times of increasing complexity.

—James Hodson
Chief executive officer, AI for Good Foundation

The Tao of Leadership is an essential guide for navigating the rapidly evolving landscape of technology and innovation with a human-centered approach. Jack Myers masterfully blends ancient wisdom with modern leadership strategies, offering profound insights for leaders facing the challenges of AI and digital transformation. This book is a must-read for anyone looking to lead with empathy, creativity, and vision in the age of machine intelligence.

—Shelly Palmer
Professor of advanced media, S. I. Newhouse School at Syracuse University

This book is a guide on how to balance, unite, and integrate the unstoppable tsunami of change driven by technology and the core humanity and wisdom that never change across the ages, a fusion that will help every individual, leader, and firm forge their future.

—Rishad Tobaccowala
Author of Rethinking Work *and former chief strategist, Publicis Groupe*

In *The Tao of Leadership*, Jack Myers provides an essential guide for leaders navigating the intricate intersection of technology and humanity. His insights on embracing creativity and empathy in the AI era resonate deeply with today's forward-thinking executives, urging us to lead with vision, courage, and compassion. This book is a blueprint for harnessing innovation while championing the human spirit.

—Kay Koplovitz
Cofounder and chairman, Springboard Enterprises

With *The Tao of Leadership*, Jack Myers offers leaders a heartfelt and essential guide to harmonizing human values with technological advancements. You'll see your leadership challenges reflected in these pages, with solutions grounded in empathy and integrity. And you'll receive the courage to implement and pursue these challenges with optimism.

—Deborah Wahl
Former SVP, global chief marketing officer, General Motors; Forbes Hall of Fame marketer

Jack Myers's *The Tao of Leadership* offers a powerful vision for future leaders, where technological innovation and human creativity harmonize to drive societal progress. We have long understood the importance of integrating cutting-edge technology with social responsibility and leadership in the academy. This book serves as an inspiring blueprint for educators and leaders alike, encouraging us to embrace empathy and ingenuity as guiding forces in shaping a more inclusive and innovative world.

—**Gracie Lawson-Borders, PhD**
Professor and dean emerita, Cathy Hughes School of Communications at Howard University

Jack Myers's *The Tao of Leadership* resonates deeply with its profound insights on leading with empathy and creativity amid technological advancements. It serves as a compelling call to action for leaders in all sectors, emphasizing the significance of human ingenuity, inclusion, and belonging as key drivers of success. Myers's perspective provides a road map for integrating innovation with the human spirit, a crucial aspect in today's landscape. A must-read for those seeking to navigate the evolving realm of leadership in the era of AI.

—**Aaron Walton**
Founder and CEO, Walton / Isaacson Advertising (WI)

In *The Tao of Leadership*, Jack Myers distills the essence of leading with both innovation and empathy—a balance that has defined the best of the media and advertising industry for decades. His insights are a master class in navigating the rapid currents of technological change while holding steadfast to the human values that are foundational to true leadership. This is an important read for leaders committed to crafting a future where technology amplifies, rather than replaces, our humanity. Jack understands that it isn't just about navigating today's big storms but also mastering the ripples before they become waves.

—David Verklin
Senior advisor, Boston Consulting Group

Jack Myers's *The Tao of Leadership* is an essential guide for managing technology and innovation from a human perspective. By blending ancient wisdom with modern leadership strategies, Jack has filtered out a lot of the noise and hyperbole surrounding AI and digital transformation.

—Mike Kerans
Principal, C.R.O. Partners, Inc.

Jack Myers's *The Tao of Leadership* provides a compelling framework for leading with purpose in the rapidly evolving landscape of corporate leadership and social impact. His emphasis on empathy, balance, and integrity is directly aligned with the priorities of today's leaders who are focused on harmonizing technological advancements with human creativity to drive meaningful impact in business and society. Jack's principles will resonate deeply with business leaders, as they seek to navigate the complexities of a transformational era where authenticity and human connection are critical to achieving sustainable outcomes.

—Mark McGinn
Executive director, Sustainability and Social Impact, Edelman

PREVIOUS BOOKS BY JACK MYERS

Adbashing: Surviving the Attacks on Advertising (1993)

Reconnecting with Customers: Building Brands and Profits in the Relationship Age (1998)

Virtual Worlds: Rewiring Your Emotional Future (2007)

Hooked Up: A New Generation's Surprising Take on Sex, Politics and Saving the World (2012)

The Future of Men: Masculinity in the Twenty-First Century (2016)

THE TAO OF LEADERSHIP

之道 领导

THE TAO OF LEADERSHIP

Harmonizing Technological Innovation and Human Creativity

IN THE AI ERA

JACK MYERS

Advantage | Books

Copyright © 2025 by Jack Myers.

All rights reserved. No part of this book may be used or reproduced in any manner whatsoever without prior written consent of the author, except as provided by the United States of America copyright law.

Published by Advantage Books, Charleston, South Carolina.
An imprint of Advantage Media.

The Tao of Leadership is a registered trademark.

ADVANTAGE is a registered trademark, and the Advantage colophon is a trademark of Advantage Media Group, Inc.

Printed in the United States of America.

10 9 8 7 6 5 4 3 2 1

ISBN: 979-8-89188-171-6 (Hardcover)
ISBN: 979-8-89188-172-3 (eBook)

Library of Congress Control Number: 2024923062

Cover design by Matthew Morse.
Layout design by Ruthie Wood.

This publication is designed to provide accurate and authoritative information in regard to the subject matter covered. It is sold with the understanding that the publisher is not engaged in rendering legal, accounting, or other professional services. If legal advice or other expert assistance is required, the services of a competent professional person should be sought.

Advantage Books is an imprint of Advantage Media Group. Advantage Media helps busy entrepreneurs, CEOs, and leaders write and publish a book to grow their business and become the authority in their field. Advantage authors comprise an exclusive community of industry professionals, idea-makers, and thought leaders. For more information go to **advantagemedia.com**.

To my dad, Dave Myers, who inspired me to write and much more.

Ronda Carnegie, truly a blessing who inspires me every day.

Our children, their spouses and partners, grandchildren, stepchildren, nieces and nephews, and future generations who I hope experience the positive future I envision.

Sandra Adirondack and Gert Myers, whose intelligence and work ethic consistently challenged me to improve.

The Taoist scholars and practitioners whose insights continue to illuminate paths for personal and collective growth.

And to the brilliant minds who have dedicated themselves to technological and creative innovation throughout history.

CONTENTS

PREFACE XIX
Tao: The Path

FOREWORD XXIII

INTRODUCTION 1

**JACK MYERS'S
BLUEPRINT FOR LEADERSHIP IN THE AI ERA** 5

PRINCIPLE I 11
Harmony—Holistic Integration of
Creativity and Technology

CHAPTER 1 13
The Synergy of Technology and Creativity

PRINCIPLE II..........37
Flexibility—Innovative Thinking
with Disciplined Leadership

CHAPTER 2..........39
The Odyssey of Innovation

CHAPTER 3..........43
Viewing Change Through the Prism of Media

CHAPTER 4..........53
Michael Milken and Financial Capitalism

PRINCIPLE III..........61
Balance—Leveraging Machine Intelligence

CHAPTER 5..........63
"Real" Science Fiction

CHAPTER 6..........77
The Advantaged and Disadvantaged

CHAPTER 7..........83
Transhumanism in 2050: A Seamless Narrative

CHAPTER 8 89
Building Portals, Not Bridges

CHAPTER 9 97
The Future of Personalized Experiences

PRINCIPLE IV 103
Simplicity—Building Resilient Organizations

CHAPTER 10 105
Introducing New Human Priorities

CHAPTER 11 119
Context-Aware AI, Inclusivity, Education, Sustainability, and Globalism

PRINCIPLE V 135
Integrity—Organizational Consolidation Across Capabilities

CHAPTER 12 137
The FusionFlow Process

CHAPTER 13 163
Future-Focused Leadership Techniques

CHAPTER 14 .... **175**
Embracing the Future

CHAPTER 15 .... **191**
Future of the Workforce

CONCLUSION **211**
Harmonizing Yin and Yang in the AI Era

APPENDIX. **217**
The Myers Blueprint for Leadership in the AI Era

ABOUT THE AUTHOR. **235**

INDEX **237**

PREFACE

Tao: The Path

This book evolves from my deepest respect and admiration for the profound wisdom embodied in Taoist philosophy and the *I Ching*. While the term "Tao" is used metaphorically to signify a "path" or "way" for leaders navigating the complexities of a time of transformative change, the core philosophy of the book draws inspiration from the timeless principles of balance, harmony, and natural flow that Taoism espouses. I acknowledge that my interpretation is but a humble reflection of these ancient concepts, adapted to address modern leadership challenges. I am grateful to the Taoist scholars and practitioners whose insights continue to illuminate paths for personal and collective growth.

The *I Ching*, also known as the *Book of Changes*, is one of the oldest Chinese classical texts and is a fundamental book of wisdom and divination in Chinese culture. It dates back over three thousand years and is traditionally attributed to the legendary figure Fuxi, King Wen of Zhou, and the Duke of Zhou. The core message of the *I Ching* revolves around the concept of change and the interplay of opposite forces, symbolized by the principles of yin and yang. These principles are embedded in the philosophy of this book and are reflected in the conflict between opposing forces of those who pursue innovation and those who resist it (see chapter 12, "The FusionFlow Process").

I titled my book *The Tao of Leadership* specifically to provide leaders of today and the future with a navigational path to success as machine intelligence shifts the fundamental nature of humanity. I trust that followers of Taoism will recognize the deep respect for the meaning of "Tao" that accompanies its use to describe the path I offer for modern leaders.

Key principles of Taoism that are reflected throughout *The Tao of Leadership* include:

Wu Wei (Non-Action): In *The Tao of Leadership*, I emphasize the importance of intuitive and adaptive leadership in the AI era. Leaders are encouraged to harness the natural flow of technological advancements and market dynamics, making decisions that align with these changes without unnecessary force. This approach reflects the principle of wu wei, advocating for a leadership style that is responsive and harmonious with the evolving environment.

Simplicity and Compassion: In *The Tao of Leadership*, I underscore the value of maintaining a clear and focused vision, free from the distractions of excessive ambition. Leaders are urged to cultivate empathy, recognizing the contributions of their teams and fostering an environment of compassion and mutual respect. This mirrors the Taoist ideal of simplicity and humility, suggesting that effective leadership in the AI era requires a grounded and modest approach.

Naturalness: My leadership philosophy emphasizes authenticity and staying true to one's values. Leaders are encouraged to develop their unique strengths and lead in a manner that is genuine and in harmony with their personal and organizational ethos. This principle aligns with Taoist naturalness, promoting a leadership style that is both authentic and in sync with the natural rhythms of the organization and the broader market.

Relativity and Paradox: This book highlights the need for leaders to navigate the complexities and paradoxes of the AI era. By embracing the interconnectedness of seemingly opposing forces—such as stability versus innovation, algorithms versus creativity, AI versus ingenuity, and assertive leadership versus empathy—leaders can make more balanced and informed decisions, see the bigger picture, and understand the interconnected nature of challenges and opportunities.

Compassion, Moderation, and Humility: Compassion, empathy, and creativity are central to *The Tao of Leadership* vision. Advocacy for a leadership style that prioritizes the well-being of employees, fosters ethical decision-making, and promotes a balanced approach to growth and innovation integrates these Taoist treasures. This leadership path offers a compassionate and ethical framework for leading in the AI era, ensuring that progress is achieved without sacrificing core human values.

These goals are embraced throughout *The Tao of Leadership,* and while aspirational, they are qualities that differentiate successful leaders in age of AI. By integrating the timeless wisdom of Taoism and the *I Ching* with contemporary leadership principles, *The Tao of Leadership* provides a holistic path for navigating the complexities of the AI era. This approach not only fosters personal growth but also cultivates a thriving, adaptive, and harmonious organizational culture.

FOREWORD

We are witnessing the end of innovation as we know it—not a cessation but a profound transformation. Change is accelerating at an unprecedented rate, advancing beyond the capacity of leaders to manage without integrating machine intelligence and generative AI into organizational systems, processes, and culture. This integration is not merely a necessity for keeping pace with change; it is also the gateway to a transformative era.

As we enter the age of AI, we also stand on the brink of a new creative renaissance. History shows that every significant technological advancement sparks a corresponding surge in creativity. AI now offers tools that expand our creative potential, enhancing human creativity, which is enriched by empathy, intuition, and ingenuity. While AI can process data and generate outputs at an unparalleled scale, it is human creativity that infuses these outputs with meaning and emotional resonance.

Technological advances now integrate into daily life at unprecedented speeds, spanning AI, biotechnology, and digital communication. Innovations that once took decades to mature now do so in a fraction of the time, pressuring organizations to shorten the timeline from ideation to implementation.

THE TAO OF LEADERSHIP

The rapid evolution of generative AI exemplifies this accelerating pace, rendering previous versions obsolete almost immediately. This rapid progression highlights the insufficiency of traditional corporate structures, characterized by rigid hierarchies and slow decision-making processes.

As we navigate this transformative era, leaders must embrace the dynamic nature of technology and foster a culture of continuous adaptation and creativity. They must be prepared to harmonize the emerging potential of AI and the unique qualities of human ingenuity, driving innovation that is both impactful and enduring.

This book explores the strategies and mindsets required to lead in such times of unparalleled change. It provides insights on balancing technological advancements with human creativity and offers a blueprint for cultivating a leadership style that is both innovative and resilient. Join me on this journey to understand how we can thrive in the age of AI and beyond, embracing a future where creativity and technology harmoniously coexist.

INTRODUCTION

Since the invention of the wheel and of fire, humans have mastered and bent technology to their control. We've now entered a stage of history that futurists and science fiction writers have long imagined—a new reality that will ensnare us all in a spectacle of human upheaval and loss of control. Identity, culture, careers, and connections are challenged as the fabric of reality is interwoven with the digital ether.

The acceleration of technology since the dawn of the internet browser in 1993 has fundamentally reshaped society, businesses, and human experience. The introduction of OpenAI in 2020 marked a pivotal moment, pushing us into an era where technological advancements outpace our capacity to control them.

The convergence of biological and machine intelligence is now redefining what it means to be human. The boundaries between the tangible and the virtual, the authentic and the artificial, are becoming increasingly blurred. In the global village, information flows at an unprecedented rate, challenging our ability to make meaningful decisions.

This book is not just about technological progress; it's also a profound reflection on humanity's journey through this transformative period. It urges us to consider our legacy, the choices that have brought us here, and the decisions that will shape our future. As we look ahead, we must contemplate the role of businesses, corporations, and leaders in sculpting our societies and defining our collective destiny.

The Tao of Leadership represents the first comprehensive blueprint for corporate leadership since Peter Drucker's 1952 manifesto, *Concept of the Corporation*, providing the navigational tools for leading organizations into a future designed for the twenty-first century.

In this new reality, the inherent inertia within corporate, governmental, and human spheres is driving a greater dependence on machine intelligence. This dependence will not only empower but also necessitate the role of intelligent machines in shaping economies, societal norms, and cultural identities.

Intelligence, as defined by dictionaries such as *Merriam-Webster* and *Britannica*, encompasses the ability to learn, understand, and adapt. This definition extends beyond humans, acknowledging the capabilities of AI. As we stand on the cusp of releasing advanced forms of generative, intuitive, empathetic, and sentient intelligence, we must prepare for a future where human interaction norms are challenged, leading us toward either transcendence or oblivion.

The period from the advent of ChatGPT in 2020 to the midpoint of the twenty-first century will define the future of our civilization and humanity's role within it. While science fiction often paints a dystopian future, this book presents a more hopeful vision. It invites us all to find our own North Star as we navigate these transformative decades.

This narrative aims not only to envision the future but also to spark a dialogue about the future we want to create. It's a call to action, urging us to approach the age of computational intelligence with awareness, intention, and a commitment to preserving the essence of humanity. As the boundaries between human and artificial cognition

INTRODUCTION

blur, institutions—from families to governments—will undergo radical transformations.

Machine intelligence can guide us toward a more enlightened and egalitarian civilization. Our journey through the twenty-first century can be a testament to the resilience of the human spirit and our inherent quest for purpose and meaning.

My perspectives and deep understanding of organizations and business models have been honed over five decades as a media ecologist, observing culture and advising many of the world's largest corporations, including General Motors, Comcast/NBCU, Microsoft, CBS, TJX Companies, Aegis/Carat, Campbell Soup, and the Walt Disney Company. I make the case that leaders must rethink the structures that underpin our corporate landscape. This book brings to the forefront the need for a critical reexamination of traditional organizational frameworks considering profound generational change, rapid technological advancements, and evolving market dynamics. *The Tao of Leadership* provides a blueprint for redefining leadership in an era where the pace of technological innovation is accelerating like never before.

The traditional phases of innovation—once marked by a gradual journey from idea to implementation—must now adapt to an era where the development cycle for technologies and business models is drastically shortened. In this new landscape, the emphasis shifts from mere technological advancement to a holistic revitalization of how human creativity is fostered within corporate environments.

Realizing the full potential of this era requires a radical rethinking of the balance between machines and humans, a reimagining of business operations, and the birth of a new age of creativity.

Traditional organizational models, with their rigid hierarchies and compartmentalized departments, are out of sync with a dynamic, interconnected, and digital world. The rise of machine intelligence necessitates a more agile, responsive, and integrated approach to business management. With the timeline from invention to market readiness drastically reduced, now is the time to redefine innovation, creativity, and the organizational models that have supported businesses for centuries.

I propose a new corporate architecture that prioritizes rapid adaptation, open communication, and collaborative problem-solving. *The Tao of Leadership* outlines practical strategies for leaders to dismantle silos, promote cross-functional teams, and build a culture of continuous learning and adaptation. It also emphasizes the importance of ethical frameworks, ensuring that business practices positively impact society and the environment. My Blueprint for Leadership offers structured guidelines for leaders to champion a shift toward a more integrated, responsive, and ethically grounded corporate structure.

Leaders are focused not only on surviving in the age of machine intelligence but also on thriving by leading with vision, empathy, creativity, and a deep commitment to reshaping the corporate world. They challenge existing paradigms and succeed by following a comprehensive road map for building organizations capable of navigating the complexities of the twenty-first century.

JACK MYERS'S BLUEPRINT FOR LEADERSHIP IN THE AI ERA

Imagine navigating through a storm. The winds of change are fierce, the waves of innovation crash unpredictably, and yet, amid this chaos, a steady hand at the helm can guide the ship to safety. This is the essence of Jack Myers's Blueprint for Leadership—a transformative strategic framework designed to balance the relentless drive for innovation with the critical need for stability.

With technological advances outstripping our ability to fully grasp their implications, businesses face a unique challenge. They must innovate continuously while maintaining a solid foundation. The Myers Blueprint offers a new theory of human and business evolution, one that seamlessly integrates advanced technological solutions and harmonizes the power of machine intelligence and human creativity to navigate this accelerating pace of change.

This blueprint is not just a philosophy; it's also a practical tool kit that provides business leaders with five core principles for integrating advanced technology with human ingenuity for transformative success.

We live in an era marked by runaway technological advancements, radical generational shifts, regulatory uncertainty, and economic volatility. Traditional management theories and ivy-covered consulting models are no longer sufficient to guide strategic planning. The Myers Blueprint provides the tools and processes needed by leaders to maintain control and coherence as they manage rapid change. It equips them to manage the complex interplay between disruptive innovation and organizational stability. By implementing this framework, leaders can proactively steer their companies through turbulent times, ensuring they seize new opportunities without compromising their core stability.

The Myers Blueprint fosters a culture that balances innovative thinking with disciplined execution, leverages machine intelligence to enhance decision-making and strategic foresight, prioritizes creativity and creators, and builds resilient organizational structures capable of adapting to continuous disruption.

Here's why this matters. The future is unpredictable, but how we prepare for it doesn't have to be. The Myers Blueprint helps organizations and their leaders prepare by focusing on five core principles, which form the basis for the sections of this book.

I. Harmony: Holistic Integration of Creativity and Technology

Success in the twenty-first century requires the harmonious integration of human creativity and technological capabilities. The Myers Blueprint stresses the importance of leveraging the unique strengths of both. Just as Taoism advocates for living in harmony with the natural world, this leadership principle suggests that technology should enhance, not replace, human creativity. It's about empowering human teams with the tools and insights provided by advanced technologies while ensuring that technology serves to advance human creativity and ingenuity and vice versa. This integrated approach maximizes the potential of both human and machine intelligence, driving innovation and growth.

II. Flexibility: Innovative Thinking with Disciplined Leadership

It's easy to get swept up in the excitement of new ideas. The Myers Blueprint teaches that true innovation requires not only creativity but also discipline. By aligning their actions with the natural flow of creativity and strategic needs, leaders can ensure sustained, strategic advantage without unnecessary force. In an age when the distance between ideation and implementation is nonexistent, this balance ensures that innovation is not just an unrealized idea but also a sustained, strategic advantage.

III. Balance: Leveraging Machine Intelligence

Leveraging machine intelligence in decision-making aligns with achieving balance between opposing forces. Leaders who achieve balance create a stable and peaceful environment, fostering sustained growth and well-being. By integrating advanced analytics with creativity and disciplined leadership with empathy, leaders gain insights and skills that help them navigate the complexities and paradoxes of the modern business landscape. This interconnected understanding allows for more informed decisions, reflecting the Taoist embrace of paradox and relativity in leadership.

IV. Simplicity: Building Resilient Organizations

Change is inevitable, but resilience is a choice. Building resilient organizations requires creating structures that can absorb shocks, adapt swiftly, and emerge stronger from disruptions while emphasizing the reduction of complexity and a focus on essential values. In leadership, simplicity fosters clarity and efficiency, allowing leaders to cut through noise and concentrate on what truly matters. This principle aligns with Taoist principles, suggesting that a straightforward, empathetic approach to organizational design can enhance resilience. By avoiding complexity and focusing on essential strengths, leaders can ensure their organizations remain stable and adaptable. The Myers Blueprint redesigns the organizational hierarchy to keep your organization afloat and thriving, no matter what the future holds.

V. Integrity: Organizational Consolidation Across Capabilities

The Myers Blueprint focuses on breaking down silos and consolidating capabilities with compassion, moderation, and empathy. By fostering better communication, reducing inefficiencies, advancing innovation, and ensuring efficient resource use, this approach creates a unified organization that is agile and effective in responding to challenges and opportunities. Leaders who prioritize inner truth and moral principles inspire trust and loyalty, creating an ethical and transparent organizational culture that reflects authenticity and sincerity—core elements of integrity.

The Myers Blueprint emphasizes a new concept of the corporation that empowers leaders to create more cohesive and agile teams, foster better communication, reduce duplication of efforts, and ensure that resources are used more efficiently.

Call to Action

The Myers Blueprint is a call to action for leaders who are ready to embrace the future with confidence and courage. It's about creating a new kind of organization—one that is agile, adaptive, unshakably stable, and dedicated to investing in human creativity.

The Myers Blueprint for Leadership is your guide to leading in the AI era. It offers a comprehensive road map for creating organizations that are ready for anything. By integrating Taoist principles such as wu wei, simplicity, naturalness, relativity, and the three treasures into the blueprint, leaders can navigate the AI era with wisdom and balance. This approach not only guides you on the path to leading with vision and fosters innovation and stability but also ensures ethical and humane leadership, creating a legacy of enduring success.

The journey begins here, and the destination is a future where your organization stands as a beacon of stability and growth.

PRINCIPLE I

Harmony—Holistic Integration of Creativity and Technology

CHAPTER 1

The Synergy of Technology and Creativity

As organizations prepare for an environment dominated by machine learning and generative AI, the role of business leaders will undergo a seismic shift. Leadership as defined by Peter Drucker in *Concept of the Corporation* is now as outdated as Stone Age construction techniques. The leadership qualities developed during the digital transformation that began a few decades ago serve only as stepping stones to the more challenging management skills demanded by the rapid acceleration of technological advancements.

Effective leadership today demands a blend of tradition and futurism, requiring leaders to be agile, informed, and visionary. As innovation outpaces human capacity to navigate the data-rich and predictive complexity of modern business, new forms of leadership are emerging. Leaders must now contend with the complexities of AI and machine intelligence, which add dimensions and layers to an already challenging reality. Beyond technical proficiency and management skills, leaders must understand and apply the crucial humanistic qualities of creativity, intuition, ingenuity, and empathy.

These qualities are essential for navigating the nuanced interplay between technology and human experience. Creativity drives innovation by envisioning possibilities beyond current constraints. Intuition allows leaders to make decisions in the face of uncertainty, guiding organizations through uncharted territory. Ingenuity inspires novel solutions to complex problems, pushing the boundaries of what is achievable. Empathy fosters a deeper connection with employees, customers, and stakeholders, ensuring that technological advancements enhance rather than undermine the human experience. This leads to an exploration of the history of technological advances that coincided with bursts in creativity.

A Historical Perspective

In the tapestry of human history, periods of rapid technological advancement have consistently coincided with extraordinary waves of creativity. These transformative times have both reshaped industry and revolutionized the arts, music, theater, media, and marketing. Understanding this synergy offers profound insights into how leaders can harness today's technological advances and foster a new explosion of creative energy.

Let's begin our journey in Florence, Italy, during the fifteenth-century Renaissance, an era recognized for an explosion of knowledge and creativity. The invention of the printing press by Johannes Gutenberg in 1440 democratized information, making books and ideas accessible to a broader audience. This technological leap fueled intellectual curiosity and artistic expression. At the same time, Leonardo da Vinci, Michelangelo, and Raphael emerged as creative titans, blending art and science in ways that continue to inspire us today. Their work was not just about beauty; it also reflected the profound shifts occurring in society, spurred by technological innovation and a renewed thirst for knowledge.

Fast-forward to the late eighteenth and early nineteenth centuries, when unprecedented advancements in manufacturing, transportation, and communication—such as the steam engine, mechanized looms, and the telegraph—transformed societies and economies. During this technological upheaval, we witnessed the Romantic movement in literature, music, and art, which emerged as a response to industrialization, emphasizing emotion, nature, and individualism. Composers such as Ludwig van Beethoven and writers such as Mary Shelley pushed the boundaries of their respective fields, reflecting the dramatic changes and complexities of their time.

The Industrial Age in the late nineteenth and early twentieth centuries introduced a wave of technological innovation with the advent of electricity, the telephone, and the automobile. These inventions revolutionized everyday life and the way people connected with one another. The era also saw the rise of the Modernist movement in art and literature. Figures such as Pablo Picasso in visual arts and James Joyce in literature broke away from traditional forms and conventions, exploring new ways of expression and perception. Their work mirrored the rapid technological changes and the new realities they brought about.

Turning our attention to the mid-twentieth century, we find a compelling intersection of technological advancement and creative innovation in the advertising industry. The early twentieth century was a time of significant change, with the rise of radio and, later, television transforming how information was disseminated. These new media platforms provided fertile ground for creativity in advertising.

THE TAO OF LEADERSHIP

Iconic figures such as David Ogilvy, Leo Burnett, Bill Bernbach, and Jerry Della Femina harnessed the power of these technologies to craft compelling narratives that captivated audiences. Ogilvy's data-driven yet creatively inspired approach set new standards in the industry, while Bernbach, Burnett, and Della Femina pushed the boundaries with bold and innovative campaigns. Their work demonstrated how technological advancements in media could be leveraged to tell stories that resonated deeply with people.

The mid-twentieth century was also defined by the Space Age, a period marked by advancements in aerospace technology and the race to explore outer space. The launch of Sputnik by the Soviet Union and the subsequent Apollo missions by the United States captured the imagination of millions. This era of exploration and innovation inspired a burst of creativity in popular culture. Science fiction literature and films flourished, with authors such as Arthur C. Clarke and filmmakers such as Stanley Kubrick envisioning futuristic worlds and technologies. The intersection of space technology and creativity not only captivated the public but also influenced scientific research and exploration.

In the late twentieth and early twenty-first centuries, the rise of the internet, social media, and mobile technology once again both disrupted business models and spurred a new wave of creativity. This period saw the emergence of digital art, electronic music, and self-published podcasts and videos. Creators such as Björk in music and Quentin Tarantino and James Cameron in film utilized new technologies to innovate and tell stories in unprecedented ways. The digital era blurred the lines between creator and consumer, enabling more people to participate in the creative process through platforms such as YouTube, Instagram, and TikTok.

Today, we stand at the brink of another transformative era driven by machine learning and generative AI. This technology is revolutionizing not just industries but also the very nature of creativity itself. AI can analyze vast amounts of data to generate insights, create content, and even compose music or write poetry. It's an era where human creativity and machine intelligence are beginning to intertwine in unprecedented ways.

In the advertising world, the current era of machine learning and generative AI is unleashing a new wave of creative genius. Visionaries such as David Droga of Droga5 (part of Accenture Song) are integrating AI to deliver hyperpersonalized, data-driven campaigns that resonate on an individual level. Digital artists such as Refik Anadol use AI to create mesmerizing visual art installations that push the boundaries of what is possible. In literature, AI is being used to cowrite novels, offering new forms of storytelling and narrative structures. In music, artists are collaborating with AI to compose pieces that blend human emotion with algorithmic precision.

Each period of rapid technological advancement has acted as a catalyst for a cultural burst in human creativity. From the Renaissance to the Industrial Revolution, from the digital age to today's AI-driven world, technology and creativity have always been intertwined. They propel each other forward, creating new possibilities and pushing the boundaries of what we can achieve.

As we navigate this current era of machine learning and generative AI, we must simultaneously harness the power of technology to expand the horizons of human creativity.

Understanding the history of these synergistic periods allows us to recognize the potential of our current technological landscape. By learning from past intersections of technology and creativity, we can better exploit today's advancements to foster a new era of creativity.

This continuous cycle of technological progress and creative flourishing holds the promise of transforming our world in ways we are only beginning to imagine.

Successful leaders are defined by their willingness and ability to learn from the lessons of the past and embrace the possibilities of the future. By fostering a culture that values both technological prowess and creative intuition, effective leaders can unlock new realms of innovation and transform the way we live, work, and create.

The synthesis of art and science, human empathy, intuition, and ingenuity is not just a historical phenomenon but also a critical reality for leaders in today's technological landscape. As AI and machine learning advance, the need for creative integration becomes more pronounced, especially among the multigenerational workforce.

In exploring the intersection of advancing technology and the rebirth of creativity through the lens of Leonardo da Vinci, Antoni Gaudí, Pablo Picasso, and Steve Jobs, we can recognize the timeless value of integrating diverse fields and perspectives. In the age of AI, we have the opportunity to build on their legacies, creating a future where technology and creativity are not at odds but are powerful allies in the quest for human advancement.

The Da Vinci Connection: Bridging the Past and Future of Technology and Creativity

To understand the intersection of technology and creativity in our current era, we turn to the quintessential Renaissance man, Leonardo da Vinci, whose life and work epitomize the harmonious blend of art, science, and innovation.

Da Vinci, born in 1452 in the small town of Vinci, Italy, is celebrated as one of history's most extraordinary polymaths. His contributions span art, science, engineering, anatomy, and architecture, making him a paragon of the Renaissance spirit. Da Vinci's ability to integrate diverse disciplines into a cohesive whole exemplifies the creative synthesis that today's technological advancements strive to achieve.

Da Vinci's masterpieces, the *Mona Lisa* and *The Last Supper*, are not just artistic triumphs; they are also the result of meticulous scientific observation and a profound understanding of human anatomy, light, and perspective. His art was deeply intertwined with his scientific inquiries. For instance, his anatomical sketches were not only artistically exquisite but also biologically accurate, informed by his dissections and studies of the human body. This integration mirrors the contemporary relationship between AI and creativity. AI's ability to analyze vast amounts of data and recognize patterns can enhance creative processes in art, music, and literature. AI-generated art, which uses algorithms to create new visual pieces, is a modern echo of da Vinci's fusion of art and science.

Da Vinci's notebooks are filled with designs for inventions far ahead of his time, including flying machines, armored vehicles, and various mechanical devices. His approach to engineering was fundamentally creative, driven by curiosity and an insatiable desire to understand the mechanics of the natural world. Similarly, today's technological advancements in AI and machine learning are driven by a creative quest to solve complex problems and improve human life. Innovations such as autonomous vehicles, advanced robotics, and AI-driven medical diagnostics reflect the same spirit of ingenuity and interdisciplinary thinking that characterized da Vinci's work.

Leonardo da Vinci serves as a compelling case study to support the theory of a synchronous intersection between advancing technology and a rebirth of creativity. His life demonstrates that true innovation arises from the confluence of diverse fields of knowledge and the application of creative thinking to scientific and technical challenges. Da Vinci's legacy is a testament to the power of interdisciplinary thinking and the seamless integration of art and science. His ability to envision and create groundbreaking works across multiple domains underscores the potential for AI to act as a catalyst for a new era of creative innovation. Da Vinci's legacy reminds us that the pursuit of knowledge and creativity is timeless.

The Industrial Age: A Symphony of Technology and Creativity

The rich interplay between technological advancement and creativity is exemplified by the Industrial Age. Spanning the late 1800s to the mid-1900s, this era was marked by groundbreaking technological innovations such as the automobile, telegraph, and cinema, which redefined society, culture, and business. Simultaneously, it witnessed an explosion of creativity in the arts and architecture, epitomized by luminaries such as Antoni Gaudí, Pablo Picasso, and Jules Verne. These technological and creative forces complemented and amplified each other.

The Industrial Age brought about profound changes in how people lived and worked. The invention of the steam engine by James Watt in the late eighteenth century set the stage for rapid industrialization. Factories began to spring up, cities grew, and the automobile revolutionized mobility. Karl Benz's creation of the first practical automobile in 1885, followed by Henry Ford's mass production techniques, not only transformed transportation but also catalyzed economic and social shifts that influenced all aspects of life, including the arts.

One of the most fascinating figures of this era, Antoni Gaudí, embodied the seamless integration of technological innovation and artistic creativity. Gaudí, an architect based in Barcelona, revolutionized the field with his imaginative and groundbreaking designs. He invented new architectural methods, such as tilted columns for support, which revolutionized the building of cathedrals while creating structures of unparalleled beauty. His masterpiece, the Sagrada Família in Barcelona, stands as a testament to his genius, combining advanced engineering with intricate artistry. Gaudí's ability to harness Industrial Age innovations in materials and construction techniques enabled him to bring his visionary ideas to life.

Gaudí's work was not just about structural innovation; it was also deeply influenced by the natural world and a desire to create harmony between nature and human-made environments. This approach mirrors the current emphasis on sustainable design and biomimicry in architecture, where technology and creativity converge to address contemporary challenges. Gaudí's legacy demonstrates how technological advancements can be harnessed to enhance artistic expression, a principle that remains relevant as we navigate the AI era.

Similarly, Pablo Picasso's artistic journey reflects the profound impact of industrialization on creative expression. Initially trained in the classical style of the Renaissance, Picasso's exposure to the rapidly changing world around him fueled his evolution toward cubism. The fragmented, abstract forms of cubism were influenced by the mechanized, industrialized society of his time, as well as by new perspectives offered by advancements in photography and cinematography. Picasso's ability to reinterpret reality through the lens of industrial innovation parallels how AI and machine learning are today redefining creative processes and pushing the boundaries of what is possible in art.

Picasso's transition from traditional to avant-garde art forms underscores a broader cultural shift during the Industrial Age, when traditional norms were questioned and reimagined. This period of artistic experimentation and innovation was facilitated by the technological advances that provided new tools and mediums for expression. The rise of cinema, for example, offered a new platform for storytelling, allowing artists and filmmakers to explore narratives in ways that were previously unimaginable. Jules Verne, another visionary of this era, utilized the possibilities of the Industrial Age to craft imaginative works of science fiction that foresaw technological advancements such as space travel and submarines.

Verne's works, *Twenty Thousand Leagues Under the Sea* and *Around the World in Eighty Days*, captivated audiences with their imaginative foresight and detailed depictions of technological wonders. His ability to weave complex technological concepts into compelling stories highlights the symbiotic relationship between technological advance and creative storytelling. Verne's influence can be seen in the way contemporary science fiction continues to inspire technological advances and vice versa.

The Industrial Age's confluence of technology and creativity offers valuable lessons for leaders today and in the future. As we advance through the early stages of another technological revolution driven by AI and machine learning, the reality of a new renaissance of creativity is inevitable. Just as Gaudí, Picasso, and Verne harnessed the innovations of their time to push the boundaries of their respective fields, today's artists, architects, and storytellers will leverage AI to explore new frontiers of creativity.

The advances in machine learning and generative AI over the next decade promise to be transformative. AI's ability to analyze vast amounts of data, recognize patterns, and generate creative content will open new possibilities for artistic expression and innovation. The integration of AI in creative processes will enhance the accessibility and democratization of art. Just as the printing press made literature accessible to a broader audience during the Renaissance, AI-powered tools will enable more people to engage in creative endeavors, breaking down barriers to entry and fostering a more inclusive cultural landscape.

The Technological and Creative Wizardry of Steve Jobs

In the annals of history, few figures stand out as quintessential innovators who have masterfully integrated technology with creativity, art, empathy, ingenuity, and marketing skills. Leonardo da Vinci set the stage for what it means to blend art and science seamlessly. Fast-forward to the twentieth and twenty-first centuries, and we encounter another luminary whose career embodies this synergy: Steve Jobs. Jobs's work at Apple not only revolutionized technology but also redefined the intersection of technological advancement and human creativity.

Jobs, born in 1955, cofounded Apple Inc. in 1976 with Steve Wozniak and Ronald Wayne. From its inception, Apple was not just about building computers; it was also about creating tools that would amplify human capabilities and change the world. Jobs's vision for Apple was rooted in his belief that technology should be intuitive, accessible, and beautifully designed—a philosophy that guided the company's innovations and set it apart from its competitors.

The Apple I and Apple II, the company's first products, were revolutionary not just for their technical specifications but also for their user-friendly design and aesthetic appeal. Jobs understood that for technology to be transformative, it needed to be embraced by users.

This required a keen sense of empathy—understanding the needs, desires, and pain points of people. It also required an artist's touch, ensuring that the technology was not just functional but also beautiful.

One of Jobs's most significant contributions to the integration of technology and creativity was the Macintosh computer, introduced in 1984. The Mac was the first personal computer to feature a graphical user interface, making it far more accessible to the average person than the text-based systems that preceded it. The Mac's design, marketing, and the iconic "1984" Super Bowl ad campaign were all testaments to Jobs's ability to blend technological innovation with creative vision. The Macintosh wasn't just a computer; it was also a statement about the future of technology and its potential to empower creativity.

Jobs's journey wasn't without its challenges. In 1985 he was ousted from Apple, leading him to found NeXT and acquire Pixar Animation Studios. At Pixar, Jobs's leadership and vision helped transform a small graphics division into a powerhouse of creativity, producing groundbreaking films such as *Toy Story*. Pixar's success underscored Jobs's belief in the power of technology to enhance storytelling and creativity, a belief that he carried back to Apple when he returned in 1997.

Upon his return, Jobs spearheaded the development of the iMac, a product that combined cutting-edge technology with bold, innovative design. The iMac's success marked the beginning of Apple's resurgence and set the stage for a series of revolutionary products: the iPod, iPhone, and iPad. Each of these devices transformed their respective industries by seamlessly integrating technology with user-centric design and functionality.

The iPhone, in particular, epitomized Jobs's philosophy. It wasn't just a phone; it was also a powerful, portable computer that redefined how people interact with technology. The iPhone combined sleek design, intuitive user interface, and powerful technology, creating a product that was both innovative and culturally relevant. Jobs's marketing genius ensured that each product launch was an event, building anticipation and excitement among consumers.

Jobs's ability to foresee the convergence of technology and various aspects of human life was unparalleled. He understood that for technology to achieve its full potential, it needed to be deeply integrated with creativity and art. This integration is evident in Apple's retail stores, which are designed to be more than just places to buy products; they are spaces that inspire creativity and foster community.

The career of Steve Jobs serves as a blueprint for how technological advances can be most successful when integrated with creativity. It also illustrates how creativity can be most innovative and culturally relevant when merged with technology. By embracing the lessons of visionaries such as Jobs, we can ensure that this new wave of technology enhances human potential, drives innovation, and enriches our cultural landscape.

The Bilbao Effect: A Symphony of Culture and Innovation

The symbiotic relationship between technological advancements and creative breakthroughs is often referred to as the Bilbao Effect. This phenomenon, where the transformative power of art and architecture revitalizes a city, serves as a profound testament to the enduring power of creativity in shaping our world. The story of Bilbao, a small industrial town in the Basque country of Spain, is a testament to the dramatic and positive impact that bold, visionary leadership can have when it integrates technology, creativity, empathy, and ingenuity.

THE TAO OF LEADERSHIP

In 1983 a catastrophic flood devastated Bilbao, leaving the city grappling with economic decline and environmental degradation. Bilbao, once a thriving industrial center, faced a grim future unless it could reinvent itself. Local leaders, recognizing the need for a radical transformation, embarked on a bold and ambitious plan to revitalize the city. Their vision was to pivot from an industrial economy to one centered around culture and tourism.

Central to this vision was the decision to approach the Guggenheim Foundation, proposing that Bilbao become the site for a new Guggenheim Museum. At the time the Guggenheim had established its iconic presence in New York and Venice, making the prospect of expanding to a small, struggling industrial town seem unlikely. However, the leaders of Bilbao, armed with a vision of transformation, managed to persuade the Guggenheim Foundation of the potential impact such a project could have.

The Guggenheim Foundation, intrigued by the challenge and opportunity, agreed to the proposal. This decision set the stage for an extraordinary partnership that would change the fate of Bilbao and the Basque region of Spain. The next pivotal decision was the choice of architect. Frank Gehry, known for his innovative and unconventional designs, was commissioned to create the new museum. Gehry's approach to architecture—characterized by his use of cutting-edge technology, daring creativity, and deep empathy for the spaces he designs—was perfectly aligned with Bilbao's aspirations.

Gehry's design for the Guggenheim Museum Bilbao was nothing short of revolutionary. The building, with its flowing, organic forms clad in titanium, resembles a ship, paying homage to Bilbao's maritime heritage while signaling a bold, futuristic vision. The museum's innovative design, blending technology and artistry, quickly became an architectural icon, drawing global admiration and attention.

CHAPTER 1

The Guggenheim Museum Bilbao opened its doors in 1997, and the impact was immediate and profound. The once-declining industrial town was transformed into a vibrant cultural hub. The museum attracted millions of visitors from around the world, sparking a tourism boom that revitalized the local economy. This dramatic transformation is a powerful example of how creativity and cultural investment can drive economic and social revitalization. The success of the Guggenheim Museum Bilbao demonstrated that investing in culture and creativity could yield significant returns, both financially and in terms of community well-being. The industrial landscape of Bilbao, once marked by pollution and decline, was replaced by a thriving cultural economy, showcasing the transformative power of visionary leadership and creative innovation.

As we stand on the cusp of the AI age, the lessons from these historical intersections are more relevant than ever. The explosion of machine learning and generative AI promises to unleash a new wave of creativity and innovation. Just as the leaders of Bilbao envisioned a new future for their city through culture and creativity, today's leaders can harness AI to drive transformative change.

The future is bright, and the synergy of technology and creativity will be at the heart of progress. The era of AI promises to unleash unprecedented creativity and innovation, transforming how we live, work, and create. By embracing the lessons of the past and applying them to our future, leaders can chart a path that harnesses the full potential of human ingenuity and technological advancement, ensuring a thriving and vibrant future for all.

A Story of Leadership from Boomers to GenAI

Emma, a hypothetical seasoned leader at the helm of a multinational tech company, exemplifies the challenges and opportunities of leading in this new era. Belonging to Generation X, Emma's leadership journey has been marked by her resilience, adaptability, and pragmatic approach to navigating technological shifts. She has witnessed the rise of the internet, the advent of mobile technology, and now the explosion of AI, shaping her ability to manage diverse teams through these transformations.

As Emma surveys her organization, she sees a mosaic of generations, each with its own identity and cultural norms. Baby boomers are retiring with decades of wisdom and experience, leaving an indelible mark on corporate culture through their work ethic, loyalty, and dedication to capitalistic business models. Gen X and Gen Y colleagues, like Emma, are pragmatic and self-reliant, balancing work-life integration and digital transformation. Millennials, the driving force behind the tech revolution, bring a collaborative spirit and a desire for meaningful work, pushing organizations toward greater innovation and flexibility.

Among the younger employees, Emma observes Gen Z's distinct characteristics as digital natives who value diversity, inclusivity, and social responsibility. They adeptly navigate digital platforms, reflecting a deep familiarity with technology. The youngest, the emerging Gen Alpha, are already interacting with AI and machine-learning tools, having grown up with smart devices and voice assistants. This digital immersion shapes their expectations and approach to work.

The unique cohort known as the NetGen, born between 1990 and 1997, provides a road map for preparing for the AI-Gen born post-2020. This post-millennial/pre-Gen Z generation, the first to grow up with the internet, adapted to the rapid flow of information and the connected world. They pioneered social media, online collaboration, and digital entrepreneurship, demonstrating how to integrate new technologies into daily life. Looking forward to when the next generation, GenAI, enters the workforce, Emma anticipates a profound shift. This generation will experience AI not merely as a tool but also as a partner in their personal and professional lives, coexisting with technology in unprecedented ways.

Leading in this new era requires a delicate balance. The educational system struggles to keep pace with technological advancements, necessitating leaders to advocate for and invest in continuous learning and development programs that bridge the gap between traditional education and the demands of an AI-driven world. Managing the tension between different generational approaches to technology and work is one of Emma's most pressing challenges. Each generation brings unique strengths and preferences, and leveraging these differences will be crucial for fostering innovation and cohesion.

Emma understands the importance of high emotional and social intelligence in navigating these complexities. In addition to managing with empathy, creativity, and intuition, emotional intelligence involves a comprehensive understanding of one's own emotions and those of others, enabling leaders to build strong relationships and foster a positive organizational culture. To prepare for the future, Emma immerses herself in resources that enhance her emotional intelligence, social intelligence, and leadership skills, reading books such as *Emotional Intelligence: Why It Can Matter More Than IQ* by Daniel Goleman and watching TED Talks such as Brené Brown's "The Power of Vulnerability."

Recognizing that not everyone finds emotional intelligence easy to learn, Emma promotes mindfulness practices, feedback loops, and emotional intelligence training programs within her organization. By embracing continuous learning, leveraging emotional intelligence, and fostering an inclusive culture, Emma believes her organization can thrive in the AI age.

For leaders to navigate generational conflicts effectively and ensure continuous learning in the AI age, it is crucial to understand and mitigate the impact of AI on jobs. AI technologies are transforming the workplace, automating tasks previously performed by humans and creating new challenges for the workforce.

AI's integration into various industries is leading to the displacement of jobs traditionally held by humans. Roles in retail, fast food, and manufacturing are becoming increasingly automated, reducing the need for human workers. To mitigate these effects, reskilling and upskilling employees are essential. Leaders must invest in continuous learning programs that equip employees with the skills needed to work alongside AI technologies, including data analysis, AI programming, and machine learning. By reskilling and upskilling, organizations can ensure their workforce remains relevant and capable of leveraging new technologies.

Creating new roles that leverage human strengths in conjunction with AI capabilities is another important strategy. While AI can handle data processing and analysis, humans are still needed to interpret results, make strategic decisions, and provide empathetic customer service. Emerging roles such as AI trainers, data ethicists, and human-AI interaction designers highlight the evolving nature of work in the AI age.

Empathy plays a critical role in navigating these changes. Leaders must recognize the human impact of technological advances and develop solutions that support displaced workers, such as career counseling, job placement services, and access to further education.

Collaboration among organizations, industries, educational institutions, and governments is necessary to create safety nets and support systems for workers affected by AI-driven changes.

The future holds an inevitable reality of worker displacement and the potential for many workers to face limited options for meaningful employment. Thought leaders and researchers, such as Andrew Yang in his book *The War on Normal People*, highlight the urgency of addressing these challenges and advocate for solutions such as universal basic income to support displaced workers.

Prioritizing the human element and developing comprehensive solutions are essential for leaders to navigate the complexities of a machine-led business world. By leading with empathy and developing comprehensive solutions, leaders can ensure their organizations and workforce are well equipped to thrive in an AI-driven world.

In adapting AI to cultural, societal, gender, ethnic, and political differences, several strategies and technologies are emerging. Context-aware AI systems that recognize and respect cultural norms and values, multilingual and multicultural AI that can process multiple languages and dialects, and personalized AI experiences that cater to individual preferences are essential for global applicability. Engaging with diverse communities in AI development ensures these technologies reflect a broader range of experiences and needs.

By understanding the historical context of da Vinci's, Gaudí's, and Jobs's interdisciplinary genius and applying those lessons to modern technological and generational challenges, leaders can navigate the complexities of the AI age. Embracing continuous learning, leveraging emotional intelligence, and fostering inclusivity will enable organizations to harness the full potential of AI while nurturing the human spirit that drives innovation and progress.

Harmonious Innovators: Integrating Technology and Creativity

As we explore the critical role of integrating advanced AI technology and creativity in leadership, it's valuable to highlight leaders who exemplify these qualities. Building on the foundations laid by da Vinci, Gaudí, Picasso, and Jobs, and phenomena such as the Bilbao Effect, we look at contemporary leaders guiding us into the AI era. These leaders are "Harmonious Innovators," those who embody the principles of *The Tao of Leadership* by seamlessly blending technology with creativity, guided by empathy and emotional intelligence.

SATYA NADELLA: TRANSFORMING MICROSOFT WITH INNOVATION AND VISION

Satya Nadella, the CEO of Microsoft, stands out as a Harmonious Innovator who has effectively leveraged technology for creative and cultural transformation. Since taking the helm in 2014, Nadella has transformed Microsoft's corporate culture, emphasizing collaboration and continuous learning. His leadership has not only reinvigorated the company's innovation but also fostered an inclusive and growth-oriented environment.

Nadella's focus on cloud computing, AI, and other advanced technologies has driven significant business success. What sets him apart is his ability to integrate these technological advancements with a vision for cultural and creative progress. By prioritizing accessibility and inclusivity in Microsoft's products, Nadella has ensured that technological advancements benefit a broader audience, reflecting a commitment to cultural and creative progress similar to that seen in Gaudí's architectural innovations and Picasso's artistic evolution.

TIM COOK: LEADING APPLE WITH INTEGRITY AND INNOVATION

Tim Cook, who succeeded Steve Jobs as CEO of Apple, has demonstrated a remarkable balance of technological investment and creative vision. Under Cook's leadership, Apple has continued to push the boundaries of innovation with products such as the Apple Watch and advancements in health technology. Cook's focus on privacy, environmental sustainability, and ethical manufacturing practices highlights his commitment to social responsibility and empathy.

Cook's leadership style is characterized by his calm, thoughtful demeanor and his ability to listen and respond to the needs of his team and customers. He has fostered a culture of openness and collaboration at Apple, encouraging creativity and ensuring that the company's technological advancements are aligned with its core values. This mirrors the approach of leaders such as Steve Jobs and Antoni Gaudí, who combined visionary technology with deep empathy and cultural sensitivity.

SUNDAR PICHAI: INNOVATING WITH PURPOSE AT GOOGLE

Sundar Pichai, the CEO of Alphabet Inc. and its subsidiary Google, exemplifies the integration of technology and creativity through harmonious innovation. Pichai has guided Google through significant advancements in AI, machine learning, and quantum computing while maintaining a strong emphasis on ethical considerations and user-centric design.

Pichai's leadership has been marked by his commitment to making technology accessible and useful to everyone, a vision that aligns with Leonardo da Vinci's quest for knowledge and innovation. By fostering an environment of inclusivity and empathy, Pichai ensures that Google's technological developments are both groundbreaking and socially responsible, much like the cultural impact of Picasso's shift to cubism or the transformative architecture of Frank Gehry's Guggenheim Museum Bilbao.

JACINDA ARDERN: LEADING NEW ZEALAND WITH VISION AND EMPATHY

Although not from the corporate world, Jacinda Ardern, the former prime minister of New Zealand, exemplifies how harmonious innovation can lead to significant technological and cultural advancements. Ardern's empathetic and inclusive approach to leadership has garnered international acclaim. Her government's handling of the COVID-19 pandemic, climate change policies, and focus on mental health and well-being reflect her commitment to empathetic governance.

Ardern has also championed technological innovation, particularly in renewable energy and digital infrastructure, aligning with her vision of a sustainable and equitable future. Her leadership demonstrates that empathy and emotional intelligence are essential for driving meaningful change and fostering a culture of innovation and creativity, akin to the holistic vision seen in the Renaissance and the Bilbao Effect.

THE PATH FORWARD: INTEGRATING TECHNOLOGY AND CREATIVITY THROUGH HARMONIOUS INNOVATION

These leaders—Satya Nadella, Tim Cook, Sundar Pichai, and Jacinda Ardern—illustrate that harmonious innovation is crucial for guiding organizations and societies through periods of technological and creative transformation. Their ability to blend technological advancements with creative vision and empathy sets a powerful example for future leaders.

As we move further into the AI era, the lessons from these leaders become increasingly relevant. The integration of technology and creativity, guided by empathy and emotional intelligence, will be essential for fostering innovation and ensuring that technological advancements benefit society. Harmonious Innovators who prioritize cultural sensitivity and social responsibility will be best positioned to navigate the complexities of the modern world and drive sustainable, inclusive progress.

The Tao of Leadership emphasizes that the path to successful leadership in the AI era lies in this integration. By blending technology with creativity, and grounding leadership in empathy and emotional intelligence, we can unlock the full potential of human ingenuity and create a future where innovation and cultural advancement go hand in hand. This holistic approach to leadership will not only drive business success but also enrich our social and cultural landscapes, ensuring a vibrant and prosperous future for all.

The Bilbao Effect, much like the innovations of da Vinci, Gaudí, Picasso, and Jobs, illustrates that creativity and technology, when combined, can lead to extraordinary outcomes. As we move forward, the synergy of these forces will continue to shape our world, driving progress and enriching our cultural and social landscapes. The future is bright, and the path forward is clear: embrace the intersection of technology and creativity, and let it guide us to new heights of innovation and prosperity.

PRINCIPLE II

Flexibility—Innovative Thinking with Disciplined Leadership

CHAPTER 2

The Odyssey of Innovation

In my 1998 book *Reconnecting with Customers: Building Brands and Profits in the Relationship Age,* I wrote, "When I look forward to 2025 my greatest hope is that I will be writing about technologies and brands that have connected every student to the internet; provided an instant link to protection for every person in danger; advanced global friendships through email 'pen pals'; broken down cultural barriers and intolerance; solved the problems of political financing; contributed to medical advances; had a positive impact upon the use of drugs, tobacco and alcohol; and significantly improved society's access to entertainment, information, education, political involvement, and public service opportunities."[1]

Depending on the prism through which you perceive our realities as we look forward again from 2025 to 2050, it's clear that technology has impacted society, culture, business, and politics, although whether for the better is subject to interpretation. I also wrote in 1998, "We have no way of forecasting the future with any degree of certainty. We

---

1 Jack Myers, *Reconnecting with Customers: Building Brands and Profits in the Relationship Age* (Spurge Ink!, 1998), 228.

can only try to draw conclusions, anticipate the reality of change, and develop a set of practices that allow us to respond with agility and speed. We do know that change will continue unabated."[2]

The Slow March of Innovation

As we look forward to 2050, most corporations, and especially large public companies, remain locked in legacy structures unchanged since 1952 that restrain agility and speed. While innovation has been the linchpin in the evolution of societies, economies, and cultures throughout history, the journey of innovation is far from a swift or straightforward path. From the invention of the wheel to the creation of the internet, no matter how profoundly each groundbreaking development may have reshaped the human experience, innovation has always faced resistance.

This inertia stems from a variety of sources: legacy institutions entrenched in the status quo, business models built around existing technologies and practices, and a human tendency to rely on experience and tradition. Throughout history, these forces have frequently frustrated the process of innovation, delaying the adoption of new ideas and technologies.

The Essence of Innovation

Innovation, in its essence, is about seeing beyond the horizon, imagining a world transformed by new ideas and technologies. It is about the relentless pursuit of progress, driven by the belief that we can create a better future through ingenuity and invention.

---

[2] Myers, *Reconnecting with Customers*, 229.

Innovations must prove not only their technological viability but also their ability to fit within or positively transform existing economic and social frameworks. The reluctance to embrace new technologies and ideas is often rooted in the risk aversion inherent in established institutions and models. Corporations, governments, and educational systems may see more threat than opportunity in disruptive innovations, opting instead to invest in the incremental improvement of existing technologies and maintenance of organizational hierarchies.

Generative AI, even in its earliest stages, can analyze vast amounts of data, identify patterns, and generate insights that surpass human capabilities. To leverage these advancements, organizations must integrate AI into their decision-making processes, allowing machines to augment and even replace human intelligence. This requires centralized leadership to oversee integration and ensure AI aligns with the organization's strategic goals.

Leaders in the AI era must understand and embrace the need to overcome inertia, resistance, and risk aversion to innovate with speed and agility. This is crucial because the rapid pace of technological advancements demands swift adaptation. Leaders who cling to outdated structures and slow decision-making processes risk falling behind. To thrive, they must cultivate a mindset that values experimentation, embraces failure as a learning opportunity, and prioritizes agility in their strategic planning. This proactive approach to innovation ensures that organizations remain competitive and can harness the full potential of AI, driving progress and sustaining relevance in a transforming landscape.

Embracing this mindset is not merely a strategic choice but also a necessity. As AI continues to evolve, the gap between those who can adapt quickly and those who cannot will widen. Leaders must champion a culture of continuous learning and flexibility, breaking free from the constraints of legacy systems. By doing so, they position their organizations to lead in the new era, leveraging AI to create a future where human creativity and machine intelligence coexist and complement each other, driving unprecedented levels of innovation and progress.

CHAPTER 3

Viewing Change Through the Prism of Media

The narrative of innovation in media captures the essence of human ingenuity, resistance, and adaptation—a story that echoes through every industry and business category. The media journey begins in the fifteenth century with Johannes Gutenberg, a German blacksmith, goldsmith, printer, and publisher. In a modest workshop, amid the clatter of metal and the scent of ink, Gutenberg's invention of the mechanical movable-type printing press around 1440 marked the onset of the Printing Revolution. This invention fundamentally transformed knowledge dissemination, communication, and society itself.

Before Gutenberg's breakthrough, books were painstakingly transcribed by hand, primarily by monks in monasteries. This laborious process, both time-consuming and costly, restricted books' accessibility to the wealthy and powerful, such as the church and royal courts. The monks, custodians of knowledge, found their roles threatened by Gutenberg's innovation. His press enabled the rapid, accurate reproduction of texts, drastically reducing time constraints and democratizing knowledge. Despite resistance from the religious monks and the broader church hierarchy—who feared losing control over religious texts and the commodification of the sacred act of copying—Gutenberg's printing press facilitated the spread of the Renaissance, the Reformation, and the Scientific Revolution. This marked the birth of modern knowledge dissemination and democratization.

As centuries passed, the hunger for faster means of communication grew. The world needed more than just the physical transportation of texts. This void was eventually filled by the invention of the telegraph in the early nineteenth century. Samuel Morse and his assistant, Alfred Vail, developed Morse code, demonstrating the telegraph's practicality and efficiency. Skepticism and resistance came from established courier systems, such as the Pony Express, which viewed it as a threat. However, the successful transmission of a message from Washington, DC, to Baltimore in 1844 marked the beginning of its widespread adoption. Companies such as Western Union expanded the telegraph network, fundamentally altering communication by enabling the instantaneous transmission of information.

The telegraph's success laid the groundwork for another transformative invention: the telephone. In the latter half of the nineteenth century, Alexander Graham Bell envisioned a device that could transmit the human voice over wires, overcoming Morse code's limitations. Despite facing technical challenges, skepticism, and legal battles, Bell successfully demonstrated the first clear transmission of spoken words in 1876. The journey from invention to public acceptance was fraught with resistance from those who viewed the telephone as a novelty. Yet persistent advocacy and the gradual building of a network showcased its transformative impact.

CHAPTER 3

The evolution of the telephone set the stage for the development of the mobile phone, spearheaded by visionaries such as Martin Cooper of Motorola and later by entrepreneurs such as Steve Jobs, who introduced the iPhone in 2007. Transitioning to mobile technology required overcoming corporate inertia and convincing consumers to adopt new ways of interacting with technology. The stories of Bell and Jobs highlight the critical role of individual entrepreneurial innovation in overcoming resistance and driving technological progress.

From the telephone, the leap to radio and television marked another significant advancement in communication. Inventors such as Lee de Forest, who developed the Audion tube, and David Sarnoff, who envisioned the potential of radio broadcasting, faced skepticism and resistance from established entities. Despite these challenges, radio and television transformed communication by enabling the wireless transmission of voice, music, and visuals, reshaping society and culture.

One of the most transformative periods in communication history arrived with the introduction of the internet, its origins dating back to the 1960s with ARPANET. But it wasn't until the late twentieth century that it began to enter the public consciousness.

In 1991 Tim Berners-Lee invented the World Wide Web, a system that utilized hypertext to link documents and enable easy navigation. The development of web browsers, such as Netscape Navigator in 1994 and Internet Explorer in 1995, revolutionized how people accessed and shared information. These browsers provided a user-friendly interface that made the internet accessible to the masses, driving rapid adoption and fundamentally altering communication, commerce, and entertainment. The internet browser facilitated a global interconnectedness, allowing for instant communication, the rise of social media, and the creation of vast digital economies. This era underscored the importance of technological accessibility and paved the way for the digital age, where information is democratized, and global interaction is seamless.

The most recent paradigm shift comes with the advent of generative AI, a technology capable of creating content and solving problems with minimal human input. This new era, marked by the integration of AI in everyday life, represents a profound transformation in human-technology interaction.

The generation born into this era, GenAI, will navigate a landscape where human and machine intelligence coexist, shaping interactions, creativity, and problem-solving approaches.

The narrative of media innovation highlights a recurring theme: the resistance of established structures to disruptive technologies and the relentless pursuit of progress by visionary individuals. Traditional corporate structures, characterized by rigid hierarchies and slow decision-making processes, struggle to keep pace with rapid technological advancements. From the printing press to generative AI, the pattern is clear: those who fail to adapt quickly are left behind, while more agile competitors seize the opportunity to innovate and thrive. As technology advances, the life cycle of innovations contracts, making rapid obsolescence inevitable. Generative AI exemplifies this trend as it continually evolves, rendering earlier versions obsolete almost immediately.

Cable TV: A Case Study in Corporate Innovation and Inertia

The history of cable television in the United States is marked by the contributions of early pioneers who saw the potential of television beyond traditional broadcast. Among the most notable were oil wildcatters such as Bob Magness, who leveraged their entrepreneurial spirit to build the infrastructure that would become the backbone of cable TV. Magness, originally a rancher and oilman, founded Tele-Communications, Inc. in 1968, which John Malone later joined as CEO, transforming it into a cable giant. Chuck Dolan, another key figure, initially worked in the cable TV industry as a salesperson before founding Sterling Manhattan Cable, the first urban cable television system in the United States. His company later became part of what is now known as Optimum. Dolan's vision included using cable to deliver a wider variety of channels, including pay television services.

Cable television's growth was spurred by the establishment of channels that offered unique content not available on broadcast TV. Some of the earliest and most influential cable networks include Home Box Office (HBO), which launched on November 8, 1972, as the first premium cable service, revolutionizing TV by offering uncut and commercial-free movies, as well as original programming. Cable News Network (CNN), founded by Ted Turner and launched on June 1, 1980, was the first twenty-four-hour news network, changing how news was consumed and paving the way for continuous news coverage. Music Television (MTV) debuted on August 1, 1981, transforming the music industry with its focus on music videos and youth-oriented programming, becoming a cultural phenomenon. Turner Broadcasting System (TBS), originally a local Atlanta TV station, became one of the first "superstations," broadcasting nationally via satellite starting in 1976, offering a mix of movies, sports, and original shows.

Traditional broadcast networks such as CBS and NBC initially resisted the rise of cable television. These networks, along with ABC, dominated the television landscape and saw cable as a potential threat to their advertising revenues and market share. Their resistance was rooted in several factors: business model inertia, regulatory challenges, and technological hesitation. Broadcast networks were heavily invested in their existing business models, which relied on advertising revenue from free-to-air broadcasts. Cable TV, with its subscription-based model, represented a significant shift that they were hesitant to embrace. The Federal Communications Commission imposed various regulations on cable TV to protect broadcast networks, including restrictions on the types of programming cable could offer and the areas it could serve. These regulations slowed the growth of cable. The technical and infrastructure investments required for cable TV were substantial, and broadcast networks were slow to invest in the necessary technology to compete effectively in the cable space.

By the time broadcast networks recognized the potential of cable TV, they had already lost significant ground to early cable operators and networks. The success of channels such as HBO, CNN, and MTV demonstrated the viability and profitability of cable television, leading to a rapid expansion of the industry throughout the 1980s and 1990s. Cable TV's ability to offer niche programming, premium content, and continuous news coverage captured the audience's attention, which was increasingly seeking alternatives to the limited offerings of broadcast television. This shift also attracted advertisers to cable networks, further eroding the market share of traditional broadcasters.

The economic, business, and societal penalties incurred by the broadcast industry because of their inertia and failure to recognize and embrace technological changes were profound. Broadcast networks, which once held an unchallenged monopoly over television content, found themselves losing viewership and advertising revenue to the burgeoning cable networks. The lack of early investment in cable infrastructure and content development meant that by the time the networks pivoted to include cable in their portfolios, the most lucrative opportunities had already been seized by the early movers.

This resulted not only in billions of dollars in lost revenue and market share but also in a far heavier cost of entry to purchase or build viable cable networks. This pattern has repeated itself and accelerated with the advent of streaming services.

Leaping forward to 2025, the broadcast industry's failure to learn from past mistakes has resulted in even greater economic consequences. Just as they were slow to adopt cable television, traditional broadcasters were late to the streaming TV business, allowing companies such as Netflix, Amazon Prime Video, and Roku to gain a significant head start. These streaming services not only captured large portions of the audience but also set new standards for content delivery, consumer choice, and viewing habits. The financial impact on legacy broadcasters has been staggering, with billions in lost revenues and increased costs as they scrambled to develop competitive streaming platforms and original content.

My Frustration: A Personal Story

The case to be made is that technological advancement, as it accelerates, will take an even heavier toll on legacy companies that adhere to their traditional models and fail to embrace and adapt to change. The timeline for technological shifts has compressed dramatically, from centuries for the adoption of the printing press to mere decades for cable television—and now only months for generative AI. As technological shifts accelerate, the risk of delays will be greater than the costs of being early adopters, radically altering an organization's decision-making hierarchy and practices.

The history of cable television—from the pioneering efforts of TV-set salesmen and oil wildcatters delivering entertainment to their workers to the rise of powerful cable networks—serves as a cautionary tale for today's media, advertising industry, and other businesses facing similar technological disruptions. Embracing change, investing in new technologies, and being willing to disrupt their own business models are crucial strategies for survival and success in an ever-accelerating technological landscape.

Reflecting on my personal experience within the broadcast and cable industry offers a vivid illustration of how technological advancements and market dynamics can reshape entire sectors—and how being an early adopter can position one at the forefront of change.

In the mid-1970s, as a sales executive with ABC-FM Radio, I found myself at the dawn of a revolution. FM radio was being introduced to the public, driven by regulatory requirements that mandated the inclusion of the FM frequency in automobiles alongside AM. Legacy AM radio station owners were granted equal rights to the FM spectrum, reaping economic benefits while I witnessed the explosion of FM radio and the rapid decline in AM valuations. It was a transformative time, and I saw firsthand the impact of technological shifts on established industries.

My journey continued from 1976 to 1983 at WCBS-TV in New York and CBS-owned television stations. In this role I was responsible for B2B marketing, retail sales, and business development. I had the task of assessing emerging business opportunities presented by early-stage technologies such as cable television, teletext, videotex, and HDTV. Growing up in Utica, New York, with just one television station—a joint affiliate for NBC and ABC—and being an early adopter of cable TV in 1964, I understood cable's potential. With cable TV, my family enjoyed a broader array of channels, including a Syracuse CBS affiliate and channels 5 (WNEW), 9 (WOR), and 11 (WPIX) from New York City. This background gave me a unique perspective on the promise of cable.

CHAPTER 3

At CBS, I wasn't just an advocate for cable; I became its evangelist. I wrote an extensive business plan recommending investment in cable TV and urged CBS-TV Network affiliates nationwide to secure local cable news rights with their cities' cable distributors. From 1979 to 1983, I built a successful new business marketing and sales department within CBS, focusing on advising advertisers on cable strategies in exchange for increased ad spending with CBS. My team introduced innovative programs such as *InfoMarketing* and *FIRST* (Framework for Insuring Retailers' Success with TV), which expanded quickly and generated successful results for both CBS and its advertisers. However, these successes incurred the enmity of those poised to benefit from maintaining the status quo and received only lukewarm support from management.

My personal tide turned when a management shift transferred my boss to another division, and my new boss, the president of CBS-TV stations, suggested, "If you like cable so fucking much, go to work for fucking cable." This blunt suggestion was a pivotal moment and prompted me to leave CBS and start my research and consulting firm, The Myers Report, in 1984.

My entrepreneurial journey did not stop there. From 1993 to 1997, my business partner, David Houle, and I formed Television Production Partners (TPP) with funding from ten leading advertisers, including General Motors and Coca-Cola. Based on our initial research, our mission evolved to explore financial models that shifted funding for network television advertising from a cost to an investment. TPP identified an innovative alternative economic model and secured $40 million in funding guarantees from the advertiser partners.

We accurately predicted the evolution of network television and projected that our model would bring windfall revenues for advertisers, greater profitability for networks, expansion of brand/network promotional partnerships, and enhanced long-term economic viability for both broadcast and cable TV networks amid inevitable competition. Despite initial enthusiasm and validation from advertisers and network programming departments, TPP represented a threat to the status quo and led network sales leaders and advertising agencies to create obstacles and disrupt the effort, ultimately causing its demise.

The long-term losses to the networks resulting from their failure to adapt and embrace change over the past half-century are staggering, likely amounting to trillions when considering the steady decline in their market capitalization. The TPP economic model, as originally outlined, envisioned the need for and incorporated long-term solutions to talent compensation issues that led to the destructive 2023 SAG-AFTRA union strike.

As technological advancements accelerate, the penalties for delay will increase, emphasizing the critical need for companies to be proactive and assess, innovate, and invest in new technologies to remain competitive. Companies that fail to adapt to technological advancements and market shifts will face significant economic and business penalties.

CHAPTER 4

Michael Milken and Financial Capitalism

Michael Milken, often referred to as the "Junk Bond King," played a pivotal role in the growth of the cable television industry during the 1980s. As a financial capitalist, Milken pioneered the use of high-yield bonds, commonly known as junk bonds, to fund ventures that traditional banks and investors deemed too risky. This innovative form of debt equity provided crucial financing for several burgeoning cable enterprises.

Milken's work at Drexel Burnham Lambert, where he was a managing director, revolutionized capital markets by demonstrating that companies with lower credit ratings could still be viable investment opportunities. He argued that these high-yield bonds could generate substantial returns for investors while providing necessary capital to companies looking to expand. His vision opened new avenues for financing, which were particularly beneficial for the rapidly growing cable television industry.

One of Milken's most notable contributions was his financing of TBS. Ted Turner, the founder of TBS, needed substantial capital to launch CNN, the first twenty-four-hour news network, and to expand his cable empire. Traditional funding sources were skeptical of Turner's ambitious plans, but Milken saw the potential. Through the issuance of junk bonds, Milken raised the funds Turner needed, which ultimately led to the successful launch of CNN and the establishment of TBS as a major cable network.

Milken's financial strategies also supported other key players in the cable industry, including John Malone and Bob Magness's Tele-Communications, Inc. cable ventures. These investments fueled the expansion of cable infrastructure, content development, and subscriber growth, contributing significantly to the cable TV boom of the 1980s and 1990s.

However, Milken's career was not without controversy. In 1989 he was indicted on charges of securities fraud and racketeering. The charges stemmed from allegations of insider trading, market manipulation, and other illicit activities. In 1990 Milken pled guilty to six felony charges and was sentenced to ten years in prison, although he served only twenty-two months. He also agreed to pay $600 million in fines and settlements.[3]

Milken's legal troubles and imprisonment were seen by many as a necessary crackdown on unethical financial practices. However, in retrospect, Milken's actions, while aggressive and, at times, legally dubious, were instrumental in reshaping the financial markets and driving economic growth. Milken's innovative use of high-yield bonds in the 1980s laid the groundwork for a new era of financial strategies

---

3   Kurt Eichenwald, "Milken Set to Pay a $600 Million Fine in Wall St. Fraud," *New York Times*, April 21, 1990, https://www.nytimes.com/1990/04/21/business/milken-set-to-pay-a-600-million-fine-in-wall-st-fraud.html.

that would drive major technological advancements in the decades to follow. His use of high-yield bonds is now a standard tool in corporate finance, and many of the innovations he introduced have been integrated into modern financial practices.

Despite his legal battles, Milken's impact on the cable industry and the broader financial markets is undeniable. His vision and willingness to take risks enabled the growth of numerous companies that might otherwise have struggled to find funding. Today, Michael Milken is recognized not only for his controversial past but also for his contributions to financial innovation and philanthropy. His story serves as a complex legacy, highlighting the thin line between visionary financial practices and legal boundaries.

Milken's story naturally leads us into the broader narrative of how venture capital has since become a critical engine for technological growth, particularly in Silicon Valley.

Venture Capitalists: An Economic Force for Innovation

The rise of the internet browser in 1993 marked the beginning of a technological revolution. Venture capital firms, recognizing the immense potential of the internet, began pouring investments into fledgling technology companies. This surge in venture capital funding was crucial in driving the growth and expansion of Silicon Valley, fostering an environment where innovation could thrive.

Venture capitalists such as John Doerr of Kleiner Perkins, Jim Breyer of Accel Partners, and Don Valentine of Sequoia Capital played pivotal roles in this transformation. These firms and individuals were instrumental in providing the financial backing for companies that would become giants in the tech industry. Kleiner Perkins, for instance, invested early in companies such as Netscape, Google, and Amazon, while Sequoia Capital was a key investor in Apple, Cisco, and Yahoo.

The reluctance of major established corporations to similarly invest in emerging technologies during this period resulted in significant business declines and an inability to compete effectively. Companies such as Kodak, Blockbuster, and Xerox, which once dominated their respective industries, failed to embrace the rapid technological changes and were unable to adapt. Kodak, for example, missed the digital photography wave, leading to its bankruptcy in 2012. Blockbuster's hesitation to transition to an online streaming model allowed Netflix to capture the market, ultimately leading to Blockbuster's demise.

In contrast, venture capital-funded start-ups were nimble and innovative, quickly capturing market share and establishing new industry standards. Firms such as Andreessen Horowitz, cofounded by Marc Andreessen and Ben Horowitz, and Peter Thiel's Founders Fund became synonymous with forward-thinking investment strategies. They provided early funding to companies such as Facebook, Airbnb, and Palantir, which have since become leaders in their fields.

The shift in investment strategies also highlighted the changing nature of risk and reward in the business world. Traditional corporations, burdened by their established business models and risk-averse cultures, struggled to pivot and innovate. Meanwhile, venture capitalists embraced risk, understanding that substantial rewards often followed bold investments in unproven technologies.

This period of technological innovation driven by venture capital not only reshaped industries but also had profound societal impact. The internet and subsequent technological advancements have transformed how we communicate, work, and live. The reluctance of established corporations to adapt and invest in these changes emphasizes the importance of agility and forward-thinking in the modern business landscape.

As I reflect on the past three decades, it's clear that the lessons from Michael Milken's financial innovations and the rise of venture capital are more relevant than ever. Companies that fail to embrace change and invest in new technologies risk obsolescence, while those that do can redefine industries and achieve unprecedented success. The story of venture capital's role in Silicon Valley is a testament to the power of vision, risk-taking, and the relentless pursuit of innovation.

While the transformative power of venture capital and technological advancements has reshaped industries and economies, it has also exposed significant shortcomings in societal investment. The relentless pursuit of profit and innovation, enabled by privilege, overshadowed investments in social responsibility, environmental sustainability, peaceful coexistence, diversity, and political cohesion. This dereliction has led to a range of societal issues that threaten the very fabric of our communities.

Capitalism as a Double-Edged Sword

The rapid technological progress brought about by venture capital–backed companies in Silicon Valley and beyond has had a double-edged impact. On one hand, it has driven unprecedented economic growth, created millions of jobs, and revolutionized how we live and work. However, this focus on technological and financial gains has often come at the expense of broader societal welfare. For example, the environmental cost of unchecked industrial expansion, the deepening socioeconomic divides, and the erosion of public trust in political institutions are all byproducts of a system heavily skewed toward profit maximization.

Corporations, in their race to stay competitive and profitable, frequently neglected responsibilities to society. The environmental degradation caused by industries like fossil fuels, the lack of investment in sustainable practices, and the minimal focus on reducing carbon footprints are stark reminders of this neglect. The social impact of technology-driven disruptions—such as job displacement due to automation, the spread of misinformation via social media platforms, and increasing economic inequality—further underscores the failures in addressing societal needs.

Prioritizing Social Responsibility

This brings me to my primary thesis: the emerging wave of techno-humanism, driven by generative AI and machine learning, necessitates a paradigm shift in how we approach investment in technology. As we stand on the cusp of another significant technological revolution, there is an urgent need to prioritize creative and responsible applications of these advances. Without a requisite investment in the humanistic aspects of technology, the foundations of society are not only threatened but could also collapse.

Historically, periods of rapid technological innovation have typically been accompanied by social upheaval. The Industrial Revolution, while bringing about significant economic progress, also led to widespread social displacement and environmental harm. Similarly, the digital revolution has led to the rise of powerful tech giants, along with the proliferation of digital divides and privacy concerns. This historical context highlights the need for a balanced approach to innovation.

Looking forward, the integration of AI and machine learning into various aspects of life offers tremendous potential for societal benefits. These technologies can enhance healthcare, improve education, streamline transportation, and address complex global challenges such as climate change.

However, achieving these benefits requires intentional and strategic investments that go beyond profitability. It demands a commitment to ethical considerations, inclusivity, and the creative application of technology to uplift humanity.

Corporate leaders and investors must recognize that the success of their ventures is intrinsically linked to the well-being of society. This recognition involves embracing a broader vision that includes social responsibility and sustainability as core components of business strategy. Companies must invest in technologies that promote environmental sustainability, such as renewable energy and carbon capture. They must also prioritize diversity and inclusion, ensuring that the benefits of technological advancements are equitably available and distributed.

Fostering peaceful coexistence and political cohesion should be integral to corporate missions. This involves supporting policies and initiatives that promote social justice, bridging economic disparities, and enhancing community resilience. By aligning their goals with these broader societal objectives, corporations can drive positive change and ensure long-term stability.

The future vision I outline calls for a symbiotic relationship between technological innovation and humanistic values. Techno-humanism envisions a world where technology amplifies human potential, creativity, and compassion. Generative AI and machine learning, when harnessed responsibly, can be powerful tools for addressing some of the most pressing challenges of our time. But this requires a shift in mindset—from viewing technology purely as a means of economic gain to seeing it as a catalyst for holistic human development.

The lessons from Michael Milken's financial innovations, the rise of venture capital, and the ongoing advancements in AI highlight the need for a balanced approach that complements the narrative of technological progress with a strong commitment to social responsibility. Corporate leaders must look beyond the exclusive capitalistic drive for profits and embrace the emergence of uniquely human qualities built on creativity, empathy, and ethical responsibility. Only by doing so can we ensure that technological advancements contribute to a thriving, equitable, and sustainable society.

Despite challenges and resistance, innovation is a common thread woven throughout the tapestry of human history. Each era's defining inventions, from the printing press to generative AI, represent milestones in the journey toward techno-humanism—a future in which technology and humanity are increasingly intertwined.

As we stand on the cusp of unprecedented technological advancements, from generative AI to quantum computing, we are also witnessing an acceleration in the innovation cycle. What once took centuries, decades, or years now unfolds in months or even weeks. This rapid pace of innovation challenges the traditional processes and structures that have historically supported and regulated new ideas and technologies.

The inertia that once hindered innovation is now juxtaposed against a world where change is both constant and rapid. This accelerated landscape requires a radical reimagining of how we innovate, adapt, and thrive. The future of innovation is not just about the speed of technological advancement but also about how societies, cultures, and businesses can navigate and harness this unprecedented pace of change for the betterment of humanity.

PRINCIPLE III

Balance—Leveraging Machine Intelligence

CHAPTER 5

"Real" Science Fiction

In 2007 I wrote in my book *Virtual Worlds: Rewiring Your Emotional Future*, "The internet as we know it today is an amoebic form of a new universe that is just beginning to be explored and developed. There is an endless quest ahead that will be very real and fascinating."[4] As we stand on the precipice of a new universe defined by the advance of machine intelligence, these words resonate more profoundly than ever. The digital landscape has transformed in ways that blur the boundaries between the real and the virtual, most evident in the seamless fusion of human and machine intelligence.

Science fiction has long served as a prophetic lens through which we can envision the future. Historically, the genre has been remarkably prescient in predicting technological advancements. Jules Verne's nineteenth-century novels, such as *From the Earth to the Moon*, envisioned space travel long before it became a reality. Arthur C. Clarke's mid-twentieth-century predictions about geostationary satellites

---

4 Jack Myers and Jerry Weinstein, *Virtual Worlds: Rewiring Your Emotional Future* (CreateSpace, 2007).

laid the groundwork for modern communication technology. These examples underscore the power of predictive thinking in shaping our technological trajectory.

The connection between science fiction and technological innovation extends to contemporary works. *Blade Runner*, based on Philip K. Dick's novel *Do Androids Dream of Electric Sheep?*, presents a future where bioengineered humans, or replicants, are indistinguishable from real humans. This narrative foresaw the complexities of AI and humanlike machines. Similarly, *The Terminator* series, with its depiction of autonomous robots, anticipated advancements in robotics and the ethical dilemmas they pose.

The concept of hyperreality, introduced by Jean Baudrillard, becomes increasingly relevant as we navigate this new world. Hyperreality describes a state where simulations of reality are more significant than reality itself, leading to a blurring of distinctions between the real and the artificial. This idea is vividly illustrated in Greta Gerwig's film *Barbie*, where the anthropomorphized dolls, Barbie and Ken, transition effortlessly between their "unreal" and "real" worlds. For the millions of fans who made *Barbie* a commercial success, there was nothing unnatural about converting these iconic dolls into lifelike characters played by Margot Robbie and Ryan Gosling. This cultural phenomenon reflects our growing comfort with hyperreal experiences, where the lines between reality and simulation are indistinguishable.

In this hyperreal world, the implications for business and technology are profound. Just as science fiction has predicted technological advancements, understanding shifts in business models and organizational structures can also be predictive. The case of General Motors (GM) serves as a prime example. In the mid-1990s, I led a consulting practice that convinced GM to centralize its media planning and buying, consolidating from seventeen separate "agencies of record" to one unified entity, GM Planworks. This shift anticipated an industry-wide trend toward the centralization of media investments, now standard practice among major national and global marketers. GM's decision to move toward electric vehicles further exemplifies how organizations can anticipate and shape future trends.

Organizational Evolution: The GM Case Study

The irony of GM's evolution is striking, considering its historical significance in shaping Peter Drucker's *Concept of the Corporation*. Drucker's seminal work, based on his early 1950s study of GM, outlined decentralization of decision-making as a key organizational principle. GM's shifts toward centralized operations, such as consolidating media planning and aggressively moving toward electric vehicles, represent a departure from Drucker's model. This centralization allows GM to streamline operations, reduce costs, and respond more effectively to market demands.

GM's centralization efforts extend to its manufacturing processes and supply chain management. For example, GM's decision to standardize vehicle platforms across multiple models has allowed for greater flexibility and efficiency in production. Additionally, GM's move toward a centralized decision-making process in its transition to electric vehicles demonstrates its commitment to innovation and sustainability.

This exploration of GM's transformation from a decentralized to a centralized entity illustrates how companies must evolve to stay relevant. By adopting a more integrated approach, GM has not only defied its historical roots but also positioned itself at the forefront of technological and organizational innovation. As we enter a new era of hyperreality, where the distinction between human and machine continues to blur, the ability to adapt and innovate will be crucial for businesses seeking to thrive in an ever-changing landscape.

The Challenge for Traditional Organizations

The danger for organizations that continue to follow Drucker's traditional formulas lies in their potential inability to adapt to rapidly changing market conditions. As the business landscape evolves, many companies remain rigidly decentralized, even within divisions and departments within those divisions. They will invariably struggle to achieve the agility and coherence required to compete effectively. The hyperreal world, characterized by rapid technological advancements and blurred boundaries between reality and simulation, demands a more integrated and flexible approach.

Most corporate leaders today remain entrenched in traditional paradigms, often resistant to change. However, by understanding the historical relationship between science fiction, theoretical exploration, and technological innovation, they can better prepare for the future. The parallels between visionary storytelling and real-world advancements are striking. Science fiction not only entertains but also challenges us to envision the extraordinary and embrace seemingly impossible challenges.

The film *Metropolis* (1927) by Fritz Lang is a foundational work that explores themes of human and machine integration. Set in a dystopian future, the film features a robot, Maschinenmensch, designed to impersonate a human woman, Maria. *Metropolis* anticipated many modern explorations of cyborgs and AI, making it a seminal work in the genre. Similarly, *Ghost in the Shell* (1995) and its live-action adaptation (2017) explore the nature of humanity and consciousness in a world where cybernetic enhancements are commonplace. These narratives provide rich explorations of the philosophical, ethical, and existential questions surrounding the merging of human and machine.

As we consider the future of a cybernetic reality, we must also address the current state of technological advancement. Virtual reality (VR), augmented reality (AR), and deepfakes are creating highly immersive simulations that often feel more real than reality itself. These technologies are preparing us for a future where the distinction between the real and the simulated is intentionally blurred. Scientific advancements in AI, robotics, and biotechnology are bringing to life concepts that once seemed purely fictional.

In this context businesses must consider the implications of attributing machinelike qualities to humans and vice versa. This reverse of anthropomorphism, or mechanomorphism, focuses on how humans can be enhanced to resemble machines in terms of behavior, efficiency, and functionality. Films such as *RoboCop* (1987) and *Upgrade* (2018) explore these themes, presenting scenarios where human consciousness is integrated with mechanical body parts, raising questions about identity and autonomy.

While dystopian narratives often highlight the worst qualities of humanity, we must also envision a positive future where the most creative, genius, and empathetic qualities of humans are expanded and integrated into machines. This vision challenges us to harness the power of technology to enhance human potential rather than diminish it.

The world of *Ex Machina* (2014), where a young programmer administers the Turing test to an intelligent humanoid robot, explores the ethical and emotional complexities of machine intelligence. Similarly, *Alita: Battle Angel* (2019) follows a cyborg's journey of self-discovery in a world of cybernetic enhancements. These stories illustrate the potential for human and machine integration to drive innovation and solve the challenges of today's digital age.

As we move toward a hyperreal world where human and machine intelligence are blended into one, we must consider the complex interplay between technological advancement, media representation, and human perception. This convergence will shape our future in ways that are both exciting and challenging.

Exploring the intricate relationship between science fiction and real-world technological advancements leads to an understanding of the predictive power of science fiction and the historical shifts in business models. Through this understanding, corporate leaders can better prepare for a future where the lines between reality and simulation are increasingly blurred. Just as science fiction has guided us through the past, it will continue to illuminate our path forward, encouraging us to dream big and embrace the extraordinary challenges ahead.

The Dawn of Cybernetic Reality

As we progress toward the third decade of the twenty-first century, dreaming big and embracing the future means looking beyond our current realities and envisioning an era I call Cybernetic Reality. This era represents a seamless integration of human intelligence and machine cognition, where authentic and artificial experiences are indistinguishable. The current state of technological advancement, media evolution, and corporate organizational models illustrates how predictions from classic and contemporary science fiction are now manifesting in our everyday lives.

The integration of human and machine intelligence is no longer a mere possibility but an emerging reality. Cybernetic Reality encompasses the science of communication and automatic control systems in both machines and living beings. It implies an environment where human minds and AI systems interact so closely that the distinction between human and machine intelligence becomes meaningless. Today's algorithm-driven decision systems are just the beginning. In this new age, decision-making will be predefined and even dictated by sophisticated programs that predict and fulfill even the most complex needs.

Howard Rheingold, in his seminal work *Virtual Reality: The Revolutionary Technology of Computer-Generated Artificial Worlds—and How It Promises to Transform Society* (1991), envisioned a future where VR would revolutionize entertainment, commerce, and interpersonal relationships. His predictions, which were perceived as science fiction when first presented, have become a reality, and their applications now extend beyond gaming and entertainment into education, healthcare, and real estate. These technologies are not just novelties; they are also rapidly advancing and will continue to transform industries and create new opportunities for engagement and innovation. VR and AR are foundations on which the rapid innovation in Cybernetic Reality is being built, inspiring real-world advancements and offering glimpses into the future.

Ray Kurzweil, in *The Singularity Is Near: When Humans Transcend Biology* (2005), predicted a merging of human and machine consciousness. He updated and accelerated his forecasts in his 2024 book *The Singularity Is Nearer: When We Merge with AI*. Singularity envisions a world where technological growth becomes uncontrollable and irreversible, resulting in unprecedented changes to society. The idea of Singularity is not merely theoretical. We witness daily advancements in AI that blur the line between human and machine capabilities. OpenAI's state-of-the-art language model has probably advanced multiple levels beyond what was available when this book was being finalized, along with multiple generative AI programs, well beyond the ability to generate humanlike text and engage in complex conversations and augment human intelligence.

Transhumanism is complementary to Singularity in its advocacy for the transformation of the human condition through advanced technologies. Transhumanism and Singularity have advanced beyond the stage of "becoming" a reality. Transhumanists are the merger of humans and machines; they enhance the traditional limitations of human intellect and physiology. Nick Bostrom, in his work "Transhumanist Values" (2005), highlights how advanced technologies elevate human creativity, empathy, and genius. This positive evolution toward creativity offers the potential for a rebirth of innovation in culture, business, and society and a solution for the polarization that is gripping society. By understanding the predictive power of science fiction and embracing the rapid innovation in cybernetics, corporate leaders can prepare for a future where the lines between reality and simulation are increasingly blurred. The decisions they make today will determine the path their organizations take, guiding them toward either success or disappearance in a future where human and machine intelligence coexist seamlessly in Cybernetic Reality.

It's essential to emphasize the relevance of embracing speculative thinking and the invaluable role of science fiction and theoretical exploration in shaping the future. For corporate managers and change agents, understanding and leveraging these imaginative frameworks is not just beneficial—it is also critical for navigating the rapidly evolving landscape of Cybernetic Reality. Speculative thinking, rooted in the rich traditions of science fiction, allows us to transcend the limitations of our current knowledge and envision futures that are both innovative and transformative. The stories of Jules Verne, Arthur C. Clarke, and Ray Kurzweil provide more than mere entertainment;

they serve as blueprints for technological advancements that once seemed fantastical but are now integral to our daily lives. By engaging with these narratives, corporate leaders can better anticipate and adapt to the disruptive innovations that lie ahead.

It's not difficult to imagine a virtual assistant that suggests movies based on your mood, curates news articles based on your immediate needs, and provides social media updates tailored to your interests and emotional state at any given moment. AI-powered smart assistants such as Amazon's Alexa and Google Assistant, inspired by science fiction's portrayal of intelligent companions, have become integral parts of daily life. These devices perform tasks ranging from setting reminders to controlling smart home systems, exemplifying the seamless blending of human and machine intelligence.

Picture yourself commuting in an autonomous vehicle on smart thoroughfares while your smart glasses overlay real-time updates about your favorite sports teams and provide immersive AR experiences. These devices will search for, find, and recommend content, learning from and anticipating your every interaction while continually refining their recommendations to ensure maximum satisfaction and positive feedback.

Advances in biotechnology are making cybernetic enhancements a reality. Cochlear implants that restore hearing to the deaf and retinal implants that provide vision to the blind are early examples of how technology will biologically embed itself to enhance human capabilities. Future developments will include advances in brain-computer interfaces that enable direct communication between the brain and external devices, further dissolving the boundaries between human and machine.

Retailers are leveraging AR to enhance the shopping experience. Companies such as IKEA offer AR apps that allow customers to visualize furniture in their homes before making a purchase. This technology not only improves customer satisfaction but also reduces return rates, demonstrating the practical benefits of AR in commerce. The rise of social media platforms has transformed how people interact, share information, and form communities. Platforms such as Facebook, TikTok, Instagram, and Pinterest have created virtual spaces where users can curate their digital personas and engage with others in real time. This digital interaction blurs the line between online and offline identities, reflecting the seamless integration of real and simulated experiences.

The proliferation of user-generated content, accompanied by high-tech editing capabilities built into mobile devices, has democratized media production, allowing individuals to become creators and influencers. This shift has given rise to new forms of expression and communication, challenging traditional media models and empowering voices previously marginalized.

As we stand on the cusp of Cybernetic Reality, we must embrace this future with open minds and a readiness to adapt. The seamless integration of human and machine intelligence will redefine our experiences, interactions, and capabilities. By understanding and harnessing these advancements, we can navigate this new era with confidence and creativity, shaping a world where the boundaries between the real and the artificial, the human and the machine, are forever blurred.

Preparing for the Future

Corporate leaders are already adapting to the shift toward a world where the boundaries between the real and the artificial, the human and the machine, are blurred by integrating advanced AI into their strategies and leveraging the next wave of predictive systems that are nearing market readiness. A proactive approach to Cybernetic Reality will become a cornerstone of successful business strategies as rapid innovation in cybernetics necessitates that organizations adapt to and integrate the advances in communication and biocontrol sciences. Leaders must embrace systems, feedback, and biometric processes that emphasize the interaction and integration of human and machine intelligence.

While many organizations have incorporated digital advancements and responded to unprecedented changes over the past few decades, they have not yet reached the point of realizing the visions once imagined in science fiction. In the words of the Canadian rock band Bachman–Turner Overdrive in their 1974 hit song, "You Ain't Seen Nothing Yet."

Despite progress, inertia within organizations remains alive and embedded. As we move toward a cybernetic world, many organizations are unprepared for a future where human and machine intelligence are interconnected to create a unified form of intelligence. Organizations are weighed down by overlapping siloed teams, each performing redundant tasks within departments, divisions, and business units. Originally intended to offset inertia, these decentralized silos now depend on it to survive. When confronted with the inability to invest in change, advocacy for the status quo becomes the norm. As Machiavelli wrote in 1513,

> It must be remembered that there is nothing more difficult to plan, more uncertain of success, nor more dangerous to manage than the creation of a new order of things. For the initiator has the enmity of all who would profit by the preservation of the old institutions, and merely lukewarm defenders in those who would gain by the new ones.[5]

---

5 Niccolò Machiavelli, *The Prince*, trans. W. K. Marriott, 1908.

THE TAO OF LEADERSHIP

Mark Zuckerberg's New Order of Things

Mark Zuckerberg, one of technology's most aggressive "initiators of a new order of things," has been victimized by the tendency toward inertia. VR and AR, bundled together and branded as the "metaverse," are prime examples of technologies that burst into the zeitgeist but gained only lukewarm defenders. The term "metaverse," first coined by Neal Stephenson in his 1992 novel *Snow Crash,* has significantly influenced contemporary discussions about the future of digital interaction and virtual environments. This concept of a VR-based successor to the internet, where users interact through avatars in a shared digital space, has inspired numerous technological advancements and strategic initiatives.

In October 2021 Zuckerberg announced a strategic rebranding of Facebook to Meta, reflecting a pivot toward building the metaverse and positioning it as the core of the company's future strategy. The metaverse, as envisioned by Zuckerberg, aims to be a fully immersive digital environment where users can work, play, socialize, and create in interconnected virtual spaces. Despite this ambitious vision, Zuckerberg's embrace of the metaverse faced significant criticism. Skeptics argued that the concept is overidealistic and that the substantial investments required may not yield the expected returns. The rebranding has been viewed as a distraction from ongoing issues within the company, such as data privacy concerns and regulatory scrutiny. Even within Meta, siloed departments remain embedded in established cultures and revenue models that depend on preserving the old institutions. Even a leader with autocratic control could not overcome the quiet and passive resistance toward change. To succeed, Zuckerberg began a slow, methodical, and structured reorganization leading toward greater centralization and integration of capabilities.

Once restructured over the next decade, Zuckerberg's vision, which remains deeply embedded in Meta's corporate culture and future strategic investments, will be operationally implemented. The company continues to pour resources into developing VR and AR technologies, which are seen as essential components of the metaverse. By investing in platforms that allow for seamless interaction between the real and virtual worlds, Meta aims to revolutionize how people engage with technology and vice versa. For instance, Meta's social VR platform, *Horizon Worlds*, allows users to create and explore virtual worlds, interact with friends, and participate in shared experiences. Additionally, Meta's investment in AI and machine learning technologies underscores its commitment to advancing the capabilities of the metaverse.

Meta's strategic focus also includes building an economy within the metaverse, where users can buy, sell, and trade digital goods and services. This vision includes creating new economic opportunities and transforming how we interact with digital content. By building interoperable platforms that allow for a seamless user experience across different applications and devices, Meta is laying the groundwork for a future where the metaverse becomes a central part of our digital lives. The adoption of the metaverse as a core strategy by Meta reflects a broader trend in the technology industry. Companies are increasingly exploring how immersive digital experiences can transform user interaction, offering new opportunities for engagement and innovation.

Addressing Realities as We Unlock New Possibilities

As organizations across various sectors begin to consider the potential of the metaverse, they will need to accelerate the breakdown of legacy silos and integrate advanced technologies, imagination, emotion, and speculative thinking into their strategic planning.

The power and importance of imagination in driving progress cannot be overstated. It is through imaginative foresight that we can identify opportunities for growth, mitigate potential risks, and inspire a culture of innovation within our organizations. Future forecasting, as an extension of this imaginative process, becomes a strategic tool for planning and decision-making. By incorporating speculative scenarios into strategic planning, businesses can cultivate resilience and agility, ensuring they remain competitive in an ever-changing market.

In the realm of Cybernetic Reality, the integration of human creativity and empathetic emotional endeavors into machine algorithms represents a profound shift. This convergence allows machines to not only perform tasks but to also engage in creative problem-solving and emotional intelligence, enhancing their utility across various domains. As we transfer these human attributes into algorithms, we unlock new possibilities for innovation and efficiency.

Globally, however, stark disparities persist. In many parts of the world, education is still denied to many, and access to the internet and mobile connectivity is limited. These disparities create a significant gap between those who can leverage technological advancements and those who cannot. The ability to master machine intelligence will increasingly define the "haves" and "have-nots."

As we prepare for this future, it is essential to address global disparities in access to education and technology. Ensuring equitable access to AI and digital technologies will be crucial for harnessing the full potential of the generation born into this new reality. By bridging digital divides and fostering inclusive innovation, we can create a society where everyone, regardless of their starting point, can contribute to and benefit from the advancements that shape our world.

CHAPTER 6

The Advantaged and Disadvantaged

In 2012 I published a research study focusing on the younger cohort of the millennial generation, those born between 1990 and 1996, a small group of 21.2 million Americans representing the transition between millennials and Generation Z. My research, documented in the book *Hooked Up: A New Generation's Surprising Take on Sex, Politics and Saving the World,* explored how this generation would impact the future. I called them the NetGen, or "N-Gen," representing both the internet and my belief that this small group would be the *engine* of an expanding economy and a positive force for society.

The N-Gen is characterized by their upbringing during the rise of the internet and digital technology, their experiences with economic challenges such as the Great Recession, and their values and behaviors shaped by significant social and cultural changes. They are known for being highly educated and tech savvy and for valuing work-life balance, social justice, and sustainability. Known for their adaptability and entrepreneurial spirit, they came of age during a period of rapid technological and social transformation. For the N-Gen, online and mobile connectivity is an integral part of everyday life. While they may reflect millennial qualities and the legacies of a pre-internet society, they relate more to their younger Gen-Z counterparts.

The Digital Divide: Amazon Versus Kodak

The advent of the internet browser in 1994 marked a significant technological breakthrough, forever altering the landscape of communication, commerce, and culture. However, this innovation also created a stark divide between the internet "haves" and "have-nots." To illustrate this divide, let's consider two hypothetical childhood friends from a small rural Midwest town, Samantha and Jake, both born in 1994.

Samantha's family, though not wealthy, prioritized technology. They purchased a computer and an internet connection soon after the release of Netscape Navigator. Samantha, curious and driven, spent hours exploring this new digital world. She taught herself coding, created websites, and connected with people from around the globe. The internet opened doors to knowledge, creativity, and opportunities that were previously unimaginable in their small town. Jake's family, on the other hand, struggled to make ends meet and viewed the internet as a luxury they couldn't afford. While Samantha delved into the digital world, Jake relied on textbooks and local resources. The lack of internet access confined his world to the physical boundaries of their town, restricting his exposure to the rapidly evolving global landscape.

As they grew older, the disparity in opportunities became more pronounced. Samantha received a scholarship to a prestigious university, thanks to her impressive online portfolio and self-taught skills.

CHAPTER 6

She thrived in an environment that valued digital literacy and innovation. Jake, despite his intelligence and hard work, found himself at a disadvantage, struggling to compete in a world increasingly dominated by technology.

The introduction of mobile devices began to bridge this gap. By the mid-2000s, mobile phones with internet capabilities became more affordable and widespread. Jake managed to save up for a smartphone, opening a new world of opportunities. With mobile connectivity, he could access educational resources, learn new skills, and connect with people beyond his immediate environment. The mobile device became a powerful equalizer, giving Jake the same level of technological access and ability as Samantha. However, Jake's early disadvantage continued to impact his ability to adapt to rapid technological advances, and he consistently followed a more traditional career path.

The corporate versions of Samantha and Jake can be seen in the stories of Amazon and Kodak. Amazon, like Samantha, embraced technology from the start. Founded by Jeff Bezos in 1994, Amazon began as an online bookstore. Recognizing the internet's potential, Bezos invested heavily in digital technology and marketing. "We are planting seeds that will grow into meaningful new businesses over time," he once said, reflecting his long-term vision. Amazon continued its proactive investment in technology with the advent of mobile devices, launching its mobile app early on and introducing the Kindle, which revolutionized the way people read books. Today, Amazon is a global e-commerce giant, continuously innovating with services such as Amazon Web Services (AWS) and Alexa.

In contrast, Kodak represents the corporate version of Jake. Once a dominant player in the photography and film industry, Kodak was slow to embrace digital technology. Despite developing the first digital camera in 1975, Kodak feared digital photography would cannibalize its film business. This reluctance was epitomized by an infamous quote attributed to Kodak executives in several articles: "We're like a deer in the headlights of a speeding train." Kodak's failure to adopt digital technology and its late attempts to pivot to digital products and services were insufficient to reverse its decline. By 2012 Kodak filed for bankruptcy, a stark contrast to Amazon's meteoric rise.

Today, Amazon continues to thrive under the leadership of Andy Jassy, who succeeded Bezos as CEO in 2021. The company remains at the forefront of innovation, leveraging advancements in AI, cloud computing, and e-commerce. Amazon's market value is surpassing $2 trillion as this book is being written, and it continues to shape the future of technology and retail.[6] Kodak, despite efforts to rebrand and focus on new technologies such as printing and packaging, has struggled to regain its former glory, illustrating how failure to adapt to technological change can lead to obsolescence.

Generational Dynamics in the Corporate World

Generational dynamics in the digital world reveal a fascinating contrast between internet natives and internet immigrants. Internet natives, such as Samantha and Amazon, grew up with the internet and mobile technology as integral parts of their lives. They are adept at navigating the digital landscape, using technology for education, work, and social interaction. Internet immigrants—those who adapted to technology later in life—often struggle to keep pace with rapid advancements. They might be proficient users but typically lack

---

[6] Associated Press, "Amazon Becomes the Fifth US Company to Reach $2 Trillion in Stock Market Value," KPBS, June 26, 2024, https://www.kpbs.org/news/economy/2024/06/26/amazon-becomes-the-5th-u-s-company-to-reach-2-trillion-in-stock-market-value.

the intuitive ease that internet natives possess. The generation born in 2020 and beyond introduces a new chapter: those native to generative AI. This AI generation will grow up in a world where machine intelligence and generative machine learning are embedded realities. For them, these technologies will be as natural and ubiquitous as the internet is to the N-Gen.

Generative AI is redefining the boundaries of what is possible. Children born into this era will experience a world where AI can create content, solve problems, and interact with humans in ways that were once the realm of science fiction. This generation will not only consume AI-driven content but also collaborate with AI in creative and intellectual pursuits. The integration of machine intelligence will enhance their learning, creativity, and problem-solving abilities, making AI a fundamental part of their cognitive tool kit.

In today's technological environment, the integration of generative AI and machine intelligence is becoming essential for corporate success. Companies that embrace these advancements, such as Amazon, thrive, while those that resist change, such as Kodak, falter. The proactive adoption of new technologies differentiates successful corporations from those that fail to adapt to the acceleration of innovation. Amazon stands as a prime example of effective AI integration. From its inception in 1994, Amazon leveraged internet technology to revolutionize e-commerce. Jeff Bezos envisioned beyond online retail, recognizing technology's potential to transform various sectors. Amazon's early AI initiatives, such as recommendation algorithms, significantly boosted customer engagement and sales. The launch of AWS in 2006 further solidified Amazon's leadership in AI, providing essential cloud computing services and tools such as SageMaker for machine learning. Additionally, Amazon's AI-powered assistant, Alexa, exemplifies the company's commitment to creating an AI-driven ecosystem, enhancing customer interaction and integrating smart home systems.

In stark contrast, Kodak's downfall highlights the risks of adhering to outdated models. Unlike Amazon, Kodak failed to invest in AI and machine learning, further entrenching its position as a "have-not" in the technology-driven economy.

CHAPTER 7

Transhumanism in 2050: A Seamless Narrative

Looking forward to the year 2050, the world has evolved into a new epoch where the lines between human and machine have blurred, heralding an age of transhumanism. The grandchildren of Samantha and Jake are now part of a generation born into this seamless integration of human and machine intelligence. Machine intelligence has become an intrinsic extension of their being. In this future Samantha's grandchildren are the torchbearers of a legacy built on the principles of creativity, empathy, and continuous learning. They lead a life where AI and machine intelligence are not just tools but also collaborators. Raised in an environment where technology is woven into the fabric of everyday life, their education has been personalized and adaptive, with AI catering to each student's unique learning style and pace. Interdisciplinary experiences begin at birth, blending technology with the humanities and fostering well-rounded individuals capable of thinking critically and creatively.

Their world is one where AI-driven assistants are commonplace, helping with daily tasks, enhancing learning experiences, and offering emotional support. These AI systems have evolved to understand and anticipate human needs, creating a harmonious coexistence where technology amplifies human capabilities. The grandchildren of Samantha lead initiatives that leverage AI to solve global challenges. They are at the forefront of projects addressing climate change, healthcare disparities, and social inequality, guided by ethical AI principles that ensure technology serves humanity's best interests.

Jake's descendants have learned from the past. Witnessing the consequences of resisting change, they have become advocates for technological adoption and innovation. By embracing new technologies and fostering a culture of continuous improvement, they have rebuilt their paths and contributed to a society that values resilience and adaptability. They play pivotal roles in championing sustainable practices and ethical business models, integrating technology in ways that enhance human well-being and environmental sustainability.

In this transformed workplace, remote work and AI-driven collaboration tools have become the norm, allowing people to work from anywhere in the world. Companies operate with flattened hierarchies, promoting a culture of inclusivity and collaboration. Humanistic values such as empathy and ethical responsibility are integral to business strategies, ensuring that technological advancements benefit society.

In 2050 the corporate landscape has shifted dramatically. Organizations are no longer confined to physical offices; instead, they operate in a virtual environment where employees interact through advanced holographic interfaces and immersive virtual reality. AI-driven systems manage routine tasks, allowing humans to focus on creative, collaborative, and strategic endeavors. The true differentiators of success—creativity, empathy, and intuition—remain uniquely human qualities.

These traits, combined with the computational power and analytic capabilities of advanced AI, drive innovation to maintain a competitive edge.

The stories of Samantha's and Jake's grandchildren offer a vision of what is possible when individuals and organizations embrace change, prioritize education, and integrate humanistic values into their core. In 2050 the world has become a place where technology and humanity coexist harmoniously, driven by a collective commitment to continuous learning, ethical responsibility, and compassionate leadership. The legacies of Samantha and Jake remind us that the choices we make today will shape the world for generations to come, guiding us toward a future where human and machine intelligence are seamlessly integrated, creating a society that thrives on creativity, empathy, and continuous innovation.

Bizarro 2050—Another Vision of the Future

In an alternate vision for the year 2050, the unchecked and rampant development of generative AI and machine learning—without the complementary influences of human ingenuity, intuition, creativity, and empathy—has led to a markedly different world. This vision, while not overtly apocalyptic, paints a stark picture of a society and corporate culture driven by a relentless pursuit of profits and technology for technology's sake, often at the expense of fulfilling basic human needs.

In this bizarro 2050, Samantha's and Jake's grandchildren live in a world where technology dominates every aspect of life, yet the absence of humanistic values has created a society marked by disconnection and inequality. Corporations, driven by the capitalistic imperative to maximize profits, have embraced AI and machine learning to an extent where human involvement is minimal. Decision-making processes are purely data driven, optimized for efficiency and profit margins, with little regard for the social and emotional well-being of employees and consumers.

The corporate landscape is characterized by highly siloed and decentralized organizations that cling to traditional economic models. The lack of a cohesive and integrated approach has led to fragmented operations, with different departments operating in isolation, often competing for profits at the expense of other divisions within the same company. The Myers Blueprint for Leadership, which emphasized breaking down silos, centralizing action units, and fostering interdisciplinary collaboration, was never applied. As a result, companies have struggled with inefficiencies, miscommunication, and consistent failures within their innovation centers.

The workplace, while technologically advanced, is devoid of the human touch. AI-driven systems manage all aspects of business operations, from hiring and firing to project management and customer service. Employees, reduced to mere cogs in the machine, experience little job satisfaction or personal fulfillment. Remote work, facilitated by advanced holographic interfaces and virtual reality, has become the norm but without the interpersonal connections and collaborative spirit that make work meaningful. The emphasis on continuous monitoring and productivity metrics has created a high-pressure environment where burnout is rampant.

Education, once a beacon of hope for fostering creativity and critical thinking, has become mechanized and standardized. AI tutors provide personalized instruction, but the curriculum is narrowly focused on technical skills and rote learning, neglecting the development of emotional intelligence, creativity, and ethical reasoning. The human element in teaching, which fosters empathy and inspiration, has been largely eliminated. Students are well versed in handling tech-

CHAPTER 7

nology but lack the soft skills necessary for holistic development and meaningful human interaction.

Society has become increasingly stratified. The gap between the technological "haves" and "have-nots" has widened, with access to advanced education and high-paying jobs concentrated among a privileged few. Those without the resources to keep up with rapid technological changes find themselves marginalized, unable to compete in an economy that values efficiency over humanity. The promise of AI to democratize opportunities has been unfulfilled, as technological advancements have primarily served to reinforce existing power structures and inequalities.

Healthcare, while technologically advanced, has become impersonal and transactional. AI-driven diagnostics and treatment plans optimize for cost and efficiency, often overlooking the nuanced needs of patients. The human aspect of care, characterized by compassion and empathy, is missing. Patients receive quick and accurate diagnoses but feel isolated and unsupported in their health journeys.

Environmental sustainability, once a priority, has taken a back seat to economic growth. AI technologies are employed to maximize resource extraction and production efficiency, often at the expense of the planet. Short-term profits are prioritized over long-term sustaability, leading to environmental degradation and resource deplet

The societal implications of this unchecked technolo advancement are profound. While AI and machine learning ha potential to solve many of humanity's challenges, their develor without the guiding principles of human creativity, empatl ethical responsibility has led to a world that is efficient but The relentless pursuit of profit has overshadowed the impo creating a society that values human well-being and fulfillr

THE TAO OF LEADERSHIP

As we look toward the future, the lessons from this alternative vision of 2050 remind us of the importance of balancing technological advancement with humanistic values. The choices we make today will determine whether we create a world where technology enhances our humanity or diminishes it. Embracing the principles of creativity, intuition, empathy, and continuous learning is essential for ensuring that the future of AI and machine learning enriches our lives and society.

CHAPTER 8

Building Portals, Not Bridges

To avoid the fate of companies such as Kodak, corporations must not only prioritize technological innovation but also adopt a new organizational blueprint. It's intuitive that companies must invest in AI, machine learning, and other disruptive technologies to remain competitive, developing AI-driven products and integrating these technologies into their operations. Complementing these investments, it's equally imperative to build portals rather than rely on legacy practices that argue for bridges from the past to the future.

The time required to envision, develop, and build bridges is inefficient. Identifying the destination and transporting organizations from atrophying businesses to growth opportunities requires transformative frameworks that enable seamless transitions into new technological paradigms, creating new pathways for innovation.

This portal approach includes rapid prototyping and iteration, emphasizing agile development cycles and continuous improvement to keep up with technological advancements. Interdisciplinary collaboration is also essential, fostering innovation across different departments and fields. Additionally, cultivating a culture that values experimentation, risk-taking, and learning from failure is vital for sustaining technological progress.

Examples of successful portal creation include Google's restructuring into Alphabet Inc., which allows it to pursue diverse technological innovations independently from its core search engine business. This restructuring enabled Google to explore AI, healthcare, and autonomous vehicles without being constrained by its original structure.

Microsoft's rapid transition under CEO Nadella is a testament to how shedding legacy organizational structures and embracing new values can lead to a successful transformation. When Nadella took over in 2014, Microsoft was primarily known for its software products such as Windows and Office. However, he recognized the need for a significant shift toward cloud computing and AI. One of Nadella's first initiatives was to instill a cultural change within Microsoft, emphasizing creativity, intuition, and empathy. He introduced a "growth mindset" philosophy, encouraging employees to embrace challenges, learn from failures, and continuously improve.

This shift promoted a culture of collaboration and learning, replacing the previous competitive atmosphere. Nadella also highlighted the importance of empathy in leadership and product development, focusing on understanding customer needs and fostering a more inclusive workplace. This approach not only led to the creation of more user-friendly products but also helped attract and retain diverse talent. By breaking down silos and encouraging open collaboration, Microsoft fostered an environment where innovative ideas could flourish.

Strategically, Nadella pivoted Microsoft to a "cloud-first, mobile-first" strategy, prioritizing the development and expansion of Azure, its cloud computing platform. Nadella's decision required innovative thinking and an intuitive understanding of the market's future needs. Azure's success is a testament to this foresight, providing scalable cloud solutions and advanced AI tools that meet the diverse needs of global businesses. This allowed Microsoft to compete directly with other cloud giants such as Amazon Web Services and Google Cloud. Azure's robust infrastructure and comprehensive services enabled businesses to migrate to the cloud, leveraging AI and machine learning tools for enhanced efficiency and innovation. Microsoft invested heavily in AI, integrating it across its product portfolio. For instance, AI enhancements in Office 365 included intelligent search, real-time language translation, and advanced data analytics. Investments in OpenAI and acquisitions such as LinkedIn and GitHub provided valuable data and platforms to further enhance Microsoft's AI capabilities.

Additionally, Microsoft transitioned from traditional software licensing models to subscription-based services, offering continuous updates and cloud-based services through products such as Office 365 and Microsoft 365. This ensured a steady revenue stream and closer customer relationships. Under Nadella's leadership, Azure grew rapidly, becoming one of the fastest-growing cloud platforms globally. Its scalable cloud solutions and advanced AI tools attracted a wide range of customers, solidifying Microsoft's position as a leader in cloud computing. Microsoft's AI initiatives also led to significant breakthroughs in various fields, including healthcare, finance, and education. For example, AI-powered tools such as the Microsoft Healthcare Bot improved patient care, while AI-driven analytics tools helped financial institutions detect fraud and manage risk effectively.

Nadella's leadership at Microsoft is marked by a profound cultural and strategic transformation that integrated creativity, intuition, empathy, and humanistic values into the organization. Empathy played a crucial role in the overhaul of Microsoft's company culture.

THE TAO OF LEADERSHIP

Nadella emphasized understanding and addressing customer needs, which led to the creation of more user-friendly products. For instance, the integration of AI into Office 365 brought about features such as real-time language translation and intelligent search, directly responding to user requirements for more efficient and accessible tools.

Another example of integrating empathy and humanistic values is the development of Microsoft's AI for Good initiatives. These projects aim to tackle societal challenges through technology. AI for Earth, for instance, applies AI to global environmental issues, helping to solve problems related to climate change, agriculture, biodiversity, and water management. This initiative underscores Microsoft's commitment to using technology for the betterment of society.

Inclusivity, a key component of Nadella's vision, is reflected in Microsoft's hiring and workplace practices. The company has focused on creating a more inclusive environment, not only to attract a diverse workforce but also to foster diverse perspectives that drive innovation. Programs such as the Autism Hiring Program highlight this commitment, aiming to bring neurodiverse talent into the organization and ensuring they have the support needed to thrive.

Microsoft's acquisition strategies also reflect Nadella's emphasis on creativity and collaboration. Acquisitions were not just about technological gain but also about integrating communities and fostering a collaborative spirit that aligns with Microsoft's broader mission. Although scaled back, Microsoft's focus on mixed reality through HoloLens demonstrates a blend of creativity and empathy. HoloLens has been used in various sectors, including healthcare, where it assists surgeons in planning complex procedures, and education, where it provides immersive learning experiences. These applications show how Microsoft leverages advanced technology to create meaningful impact on people's lives.

Microsoft's transformation under Nadella showcases how integrating creativity, intuition, empathy, and humanistic values can drive both cultural and business success. This holistic approach has revitalized Microsoft, establishing it as a technology leader that not only innovates but also contributes positively to society.

The examples of Google and Microsoft highlight the importance of embracing generative AI and machine intelligence. Companies that invest in these technologies and adopt new organizational models will thrive, while those clinging to legacy frameworks will fall behind. Recognizing the need for portals—transformative pathways that facilitate the seamless integration of new technologies—enables corporations to navigate the rapidly changing technological landscape and secure their place in the future economy.

Integrating Data, Creativity, and Empathy in Decision-Making

Effective leaders in the AI era will understand and embrace the transformative potential of combining analytics with human-centric values.

As organizations harness the power of advanced analytic tools, leaders must adapt quickly to leverage these technologies effectively while integrating creativity and empathy into their decision-making processes. Over the past three decades, leaders of most large organizations have adapted to the increasing availability of big data and advanced analytics tools. This has not always involved embracing a data-driven culture within their organizations, where decisions are grounded in empirical evidence and strategic insights. The increasing availability of big data presents leaders with an unparalleled opportunity to gain insights and drive strategic initiatives. However, to fully realize this potential, leaders must also develop the necessary skills and infrastructure to support advanced analytics. This involves investing not only in technology and training but also in cultural change.

The following steps outline how leaders can effectively integrate advanced generative data analytics into their decision-making processes:

1. Invest in Data Literacy Training: Leaders should prioritize training their teams in data literacy, ensuring that employees at all levels understand how to interpret and utilize data effectively. This includes providing access to educational resources, workshops, and certification programs that enhance data skills (see chapter 14).

2. Adopt Advanced Analytics Tools: Investing in tools that enable the collection, analysis, and visualization of data is critical. Platforms such as Tableau, Power BI, and Google Analytics offer robust capabilities for transforming raw data into actionable insights. They represent just the cusp of generative machine-based analytics that will facilitate real-time data analysis and enable leaders to make timely and informed decisions.

3. Foster a Data-Driven Culture: Creating a culture that values data-driven decision-making requires strong leadership commitment. Empirical, data-informed evidence should be accompanied by clear organization-wide expectations for data use, the celebration of data-driven successes, and ongoing support for data initiatives.

4. Integrate AI and Machine Intelligence: Leveraging AI and machine learning technologies can enhance the accuracy and depth of data analysis. These technologies can identify complex patterns, generate predictive models, and automate routine data tasks. By integrating AI into

analytics processes, leaders can uncover deeper insights and drive more strategic outcomes.

5. Promote Creativity and Innovation: Encouraging a creative approach to data analytics involves empowering teams to explore new ideas and experiment with different analytical methods. Innovation labs, hackathons, and collaborative brainstorming sessions can foster creativity, enabling teams to develop unique solutions and insights.

6. Ensure Ethical Data Use: Ethical considerations are paramount in data-driven decision-making. Leaders must establish clear guidelines for data privacy, security, and ethical use. This includes obtaining informed consent, protecting sensitive information, and ensuring transparency in how data is collected and analyzed.

7. Implement Empathetic Analytics: Incorporating empathy into analytics involves understanding the human context behind the data. Leaders should consider the broader implications of their decisions, striving to make choices that benefit employees, customers, and society. This approach enhances trust and fosters a more inclusive business environment.

The true potential of data and analytics lies not just in the ability to process and analyze information but also in the capacity to integrate creativity and innovation into these processes. Data-driven decision-making should not be confined to rigid, algorithmic outputs; it must be complemented by human intuition, creativity, and empathy. Leaders who recognize this symbiotic relationship will be better equipped to navigate the complexities of the modern business environment.

THE TAO OF LEADERSHIP

Empathy plays a crucial role in data-driven decision-making. While data provides valuable insights, it is empathy that enables leaders to understand the human implications of their decisions. This involves considering the impact on employees, customers, and broader society. By integrating empathy into analytics, leaders can ensure that their decisions are not only data informed but also aligned with the values and needs of their stakeholders. This approach fosters trust, engagement, and long-term sustainability.

Leaders should also recognize that data-driven decision-making is an ongoing process. As new data becomes available and technologies evolve, organizations must continuously adapt and refine their analytics strategies. This requires a commitment to lifelong learning, agility, and a willingness to embrace change.

CHAPTER 9

The Future of Personalized Experiences

Personalized customer experiences are becoming essential. The integration of machine intelligence and generative AI offers unprecedented opportunities to understand and meet individual customer needs. As outlined in the previous chapter, this technological advancement must be complemented by empathy and sensitivity to ensure that personalization is meaningful and respectful. Leaders must guide their organizations toward a customer-centric approach that leverages AI while addressing privacy concerns and ethical considerations.

As AI and machine learning continue to advance, they enable businesses to gather and analyze vast amounts of customer data. This data provides insights into customer preferences, behaviors, and needs, allowing organizations to tailor their offerings and interactions. Machine intelligence can process this information at a scale and speed that humans cannot match, making it possible to deliver highly personalized experiences in real time.

Integrating AI, Empathy, and Sensitivity in a Customer-Centric Approach

The potential of AI in personalization comes with significant responsibilities. Privacy concerns are paramount, and organizations must ensure that they handle customer data with the utmost care and transparency. Customers are becoming increasingly aware of their privacy rights and are more likely to trust businesses that demonstrate a commitment to protecting their information. Leaders must navigate these concerns by implementing robust data protection measures and fostering a culture of transparency and accountability.

Empathy is a critical component in the application of AI for personalized customer experiences. While AI can provide data-driven insights, it is empathy that allows businesses to understand the emotional context behind customer interactions. Integrating empathetic capabilities into AI systems can help organizations deliver more meaningful and humanlike interactions. This involves programming AI to recognize and respond to emotional cues, ensuring that customer interactions feel genuine and considerate.

To achieve this, leaders must adopt a comprehensive approach that combines AI technology with human-centric values. Most existing advanced customer relationship management (CRM) systems already integrate AI and machine learning capabilities. These systems analyze customer data to identify patterns and trends, providing valuable insights into customer preferences and behaviors. The current generation of CRM tools automate personalized communications and tailor marketing campaigns to help businesses segment customers based on their purchase history and engagement levels, enabling targeted messaging that resonate with individual needs in real time. These tools can also predict future behaviors and preferences, allowing businesses to proactively anticipate customer needs. Predictive analytics forecast demand, personalize product recommendations, and optimize pricing strategies, enhancing satisfaction and loyalty.

The next generation of AI programming with sentient, empathetic capabilities now in development involves AI systems that can recognize and respond to emotional cues. This requires training the AI algorithms to understand tone, sentiment, and context in customer interactions. Ensuring that AI interactions are designed to be empathetic and supportive will make customer service chatbots more effective, providing reassurance during stressful situations.

Tackling Privacy and Ethical Concerns

Addressing privacy concerns and ethical considerations is crucial in this process. Implementing strict data protection measures to safeguard customer information, including encryption, access controls, and regular security audits, is essential. Transparency about data collection practices and obtaining explicit consent from customers builds trust. Clearly communicating how their data will be used, along with the benefits of personalization, is vital. Developing ethical guidelines for AI use ensures that AI applications align with customer values and societal norms, and establishing an ethics committee and regulations to oversee AI initiatives can address potential ethical dilemmas.

Implementing a Human Touch

Balancing automation with a human touch is also essential. While AI can automate many aspects of customer interaction, maintaining a human touch in critical areas is necessary. Complex or emotionally charged issues may require human intervention to provide the necessary support and understanding. Using AI to enhance human capabilities rather than replace them can assist customer service representatives by providing real-time insights and suggestions, enabling them to offer more personalized and effective support.

Fostering a culture of empathy and customer centricity also requires training employees to prioritize empathy in customer interactions. Encouraging active listening and understanding creates meaningful connections with customers. Promoting a customer-centric mindset across the organization ensures that all departments, from marketing to product development, focus on delivering value to customers. Celebrating successes in customer satisfaction and sharing stories of positive customer experiences reinforces the importance of empathy and customer centricity.

Challenges and Potential of AI-Driven Personalization

The integration of AI and empathy in personalized customer experiences offers immense potential but also presents challenges. Leaders must be aware of the inherent dangers and pushback that can arise from the use of AI in customer interactions. Concerns about data privacy, the loss of human touch, and the ethical implications of AI are valid and must be addressed proactively.

Effective leadership in this context requires a balanced approach that combines technological innovation with human values. Leaders must guide their organizations in developing AI systems that enhance customer experiences without compromising privacy or ethical standards. This involves continuous monitoring and evaluation of AI applications to ensure they align with customer expectations and societal norms.

The future of personalized customer experiences lies in the seamless integration of AI, empathy, and sensitivity to individual needs. Leaders must adapt quickly to this new landscape, leveraging advanced technologies to understand and meet customer expectations while addressing privacy and ethical concerns. By adopting a customer-centric approach that prioritizes empathy and human values, organizations can deliver personalized experiences that resonate with customers and drive lasting success. The time to embrace this transformative potential is now, as the integration of AI and empathy will define the future of customer engagement and business excellence.

PRINCIPLE IV

Simplicity—Building Resilient Organizations

CHAPTER 10

Introducing New Human Priorities

I n my hopeful vision of 2050 outlined in chapter 7, the business landscape will have undergone a profound transformation, shifting from legacy corporate forces that often dehumanized business practices to a new paradigm where sentient machine intelligence prioritizes the integration of positive humanistic forces. This rapid technological evolution underscores the need for a reinven tion of human roles within corporate structures. As machine learnin and AI handle more data-driven transactions and technologic implementations, the unique human capacity for creativity, empar ingenuity, and intuition becomes a critical asset.

The future of corporate innovation hinges not on the ability to auto but on the capacity of leaders to creatively integrate new technologies ir that enhance human well-being and drive sustainable growth.

Embracing Creativity in Corporate Culture

Companies must now pivot from structures that prioritize efficiency and standardization to ones that invest equally in creativity. This shift isn't merely structural but also cultural, requiring organizations to nurture environments where creative solutions are valued. For instance, advertising investment firm Publicis Media addresses the need for upskilling because of the fast pace of marketing technology development. They have acknowledged the widespread knowledge gaps at various levels in the media industry and emphasize the importance of fresh thinking, flexibility, and innovative approaches to learning through initiatives such as the Publicis Media IQ Academy.

This academy serves as a model for the future, illustrating how organizations can bridge knowledge gaps between generations. By offering continuous learning opportunities and fostering a culture of innovation, companies can ensure their workforce remains agile and capable of leveraging new technologies effectively. Publicis Media's approach highlights the importance of integrating creative ideation with technological deployment, ensuring these tools serve broader and ethical goals.

Apple Inc. stands out as another exemplary model of integrating creativity with technological advancement. Apple is renowned for its unique ability to combine technology with creative design, setting a standard-bearer for innovation in consumer electronics. By controlling the entire product ecosystem, Apple integrates hardware, software, and services seamlessly, driving user loyalty and operational efficiency. Apple's forward-looking leadership and functionality demonstrates its capacity to innovate such as the recent push toward augmented reality. By centralizing its organization and resources, Apple moves quickly from invention to adoption to market acceptance, ensuring that its innovations reach consumers rapidly and effectively.

As we enter this new era, the role of human creativity in organizations is not diminishing but gaining new importance. Creativity becomes the driving force behind leveraging technological advancements in meaningful ways. It's not enough to adopt new technologies quickly; companies must incorporate creative ideation in their deployment, ensuring these tools serve broader societal and ethical goals. As companies and industries adjust to the accelerated pace of technological advance, the integration of creativity into corporate DNA becomes imperative.

Creativity, as a foundation of corporate culture, remains constant, while technology, by its very nature, is disruptive. This evolution will require bold rethinking of roles, responsibilities, and rewards within organizations. Publicis Media's proactive stance on upskilling and continuous learning and Apple's centralized creativity-focused structure exemplify how companies can prepare their workforce for the future, ensuring that employees are equipped not only with technical skills but also with the creative and empathetic capabilities necessary for innovation.

Organizations such as Microsoft have shown how integrating humanistic values into their core can drive success. Under Satya Nadella's leadership, as outlined under Principle II, Microsoft has emphasized empathy, inclusivity, and continuous learning, transforming its culture and business model to thrive in the age of AI and cloud computing. This holistic approach ensures that technological advancements align with human needs and ethical standards, fostering a more inclusive and sustainable business environment.

Embracing the interplay between human ingenuity and technological capability defines the next horizon of corporate and societal advancement. As we look forward, it's clear that the synergy between human creativity and technological invention will shape the future of our global society, marking a new definition for the role of leaders in the age of machine dominance.

By 2050, if Samantha's and Jake's descendants are to be navigating a world where AI and human creativity coexist harmoniously, today's leaders must invest in leveraging both, guided by principles of ethical AI and a commitment to humanistic values. Effective leadership will be guided by underscoring the importance of integrating empathy, intuition, and creativity into technological advancements, ensuring these innovations serve the greater good of employees, consumers, partners, society, the environment, profitability, and even competitors who pursue similar positive goals.

The "N-Gen" vision of their world, not as apocalyptic and dystopian but as one progressing toward inclusivity, equality, and humanism, has the potential to become a reality. The blend of sentient machine intelligence with human creativity, empathy, and intuition shifts the collective vision from fear of the unknown to excitement for the possibilities. The integration of advanced technology with humanistic values can pave the way toward a balanced and hopeful future, where the corporate world thrives on both technological innovation and human compassion.

Visionary Leaders Move Beyond Innovation to Ideation

Visionary leaders, such as Satya Nadella of Microsoft, Elon Musk of Tesla and SpaceX, Sundar Pichai of Alphabet Inc., and Arthur Sadoun of Publicis Groupe exemplify the transformative impact of having a clear and forward-thinking vision. These leaders have not only anticipated future trends and technological advancements but have also understood their potential impact on business models, customer behaviors, and industry landscapes.

Under Nadella's leadership, Microsoft has embraced AI and cloud computing, positioning itself as a leader in AI-driven solutions with its "intelligent cloud and intelligent edge" strategy. Similarly, Musk's ambitious vision for autonomous vehicles and interplanetary travel has set new benchmarks in innovation. Pichai's focus on AI-first strategies has placed Google at the forefront of AI research and development.

Visionary thinking allows leaders to foresee potential disruptions and opportunities, fostering a culture of innovation that leverages AI to gain competitive advantages and transform business operations. This culture is crucial for aligning the entire organization around common goals and ensuring successful AI implementation through cross-functional collaboration. Furthermore, visionary leaders inspire trust and engagement among employees, stakeholders, and customers, rallying support and enthusiasm, which are critical for driving change and overcoming resistance.

Conversely, the absence of visionary leadership can lead to severe consequences. Organizations that lack a forward-thinking approach risk stagnation and falling behind competitors who embrace a future-forward approach. This can result in missed opportunities, increased risk, and uncertainty as leaders fail to navigate the complexities of the future effectively. Lack of a clear and inspiring vision can lead to low employee morale and engagement, hindering an organization's overall success.

Visionary leadership is not just an abstract ideal but also a practical necessity, requiring leaders to navigate complexity and accelerate their processes for ideation, creation, and innovation. The concept of moving quickly from ideation to implementation in decision-making reflects a transformative shift in business leadership.

While innovation typically focuses on refining existing products, services, and processes, creation involves a radical rethinking of what is possible, leading to the development of entirely new paradigms or the dramatic transformation of current models. This shift is crucial in an era where machine learning and AI not only accelerate the pace of change but also create complexities that traditional corporate hierarchies—often rigid, slow moving, and locked into legacy thinking—struggle to address effectively.

As AI and machine learning advance beyond the capacity of corporate culture to adapt and integrate, they will enable systems to learn and improve autonomously, outpacing the decision-making capabilities of hierarchical organizations. The necessity for a leadership style that transcends traditional innovation becomes evident. Such leadership emphasizes ideas and creativity, allowing leaders to visualize futures that break away from current limitations. It prioritizes intuition, empowering leaders to make swift, informed decisions even when data is incomplete or the path forward is unclear.

The quality of ingenuity is essential, as it fosters resourcefulness and cleverness, particularly in navigating the novel challenges that arise when venturing into unexplored business territories. Alongside these qualities, empathy plays a pivotal role. As technology advances, it is imperative that its uses enhance human lives and address societal needs, ensuring that technological progress does not occur in a vacuum but rather in tandem with humanistic considerations.

By embodying creativity, intuition, ingenuity, and empathy, leaders can effectively steer their organizations through a landscape being defined and shaped by machine intelligence and generative AI. This approach to leadership infuses the decision-making process with unique human qualities, enabling leaders to keep pace with technological advances and set new standards.

It drives their organizations toward a future where they lead rather than simply adapt to change, creating new opportunities and models that redefine their industries. This aligns organizational goals and builds employee and customer trust and engagement. As AI continues to reshape the business environment, maintaining a visionary approach is crucial for long-term success, ensuring organizations are well prepared to thrive amid ongoing changes.

How to Develop Visionary Leadership Skills

- Continuous Learning and Curiosity: Leaders should commit to lifelong learning and stay abreast of technological advancements and industry trends. Attending workshops, seminars, and online courses can help leaders understand emerging technologies such as AI and their potential impact.

- Strategic Thinking: Visionary leaders are adept at looking beyond the day-to-day operations and envisioning the long-term future of their organizations. Developing this skill involves understanding the broader industry context, identifying potential opportunities and threats, and crafting strategic plans that are both visionary and achievable.

- Emotional Intelligence: High emotional intelligence helps leaders manage their own emotions and those of others, enabling them to foster strong relationships and build trust. Emotional intelligence can be enhanced through self-awareness practices, feedback from others, and training in empathy and conflict resolution.

- Risk Management: Transforming from risk averse to risk management involves understanding that not all risks are detrimental. Leaders can develop skills to evaluate risks more strategically, distinguishing between growth opportunities and genuine threats.

- Fostering Creativity: Encouraging a culture of creativity within the organization allows leaders and their teams to experiment and develop creative solutions without fear of failure. This could involve setting up innovation labs, hackathons, or dedicated times for brainstorming and experimenting.

- Collaborative Leadership: Working collaboratively across teams and with external partners can enhance a leader's perspective and spark ideas. Building a network of diverse thinkers can provide new insights and drive visionary thinking.

Visionary leaders possess a unique set of qualities that enable them to steer their organizations through the complexities and dynamics of modern business landscapes. They are forward-thinking, constantly anticipating future trends and challenges that may impact their industries. This ability to look ahead is complemented by their inspirational nature; they have a knack for motivating others to pursue common goals with great enthusiasm and commitment. Visionary leaders are also characterized by their resilience; they can persevere through setbacks and maintain a steady focus on long-term objectives, demonstrating a steadfastness that inspires confidence among their teams.

Adaptability is another crucial trait, as they remain flexible and responsive to changes and challenges, understanding that the ability to pivot is often necessary in a rapidly evolving environment. They are decisive, making firm and informed decisions that balance data-driven insights with intuitive understanding. This blend of foresight, inspiration, resilience, adaptability, and decisiveness equips visionary leaders to guide their organizations successfully into the future.

When Satya Nadella took over as CEO of Microsoft, he radically shifted the company's focus from traditional software products to pioneering areas such as cloud computing and AI technologies, fostering a "learn-it-all" rather than a "know-it-all" culture.[7] This demanded not only technological innovation but also a profound transformation in organizational mindset. Similarly, Anne Mulcahy, former CEO of Xerox, is credited with pulling the company back from the brink of bankruptcy by intensifying its focus on research and development and venturing into new business areas such as digital printing—a bold departure from Xerox's traditional business models.[8]

Jeff Bezos, who left a secure position on Wall Street to launch a start-up in the then-emerging field of online retail, epitomizes the transition from being risk averse to becoming a visionary leader. His relentless expansion into new ventures such as AWS, Kindle, and space travel with Blue Origin showcases a continuous pursuit of visionary goals. Together, these leaders exemplify visionary leadership by embracing change, fostering a culture of creativity, and focusing on long-term, transformative objectives.

Arthur Sadoun, as the CEO of Publicis Groupe, exemplifies visionary leadership through his relentless focus on transformation and innovation in the advertising and media landscape. Under his guidance, Publicis Groupe has successfully navigated the complexities of the digital era, pioneering the integration of data, creativity, and

---

7 Jyoti Mann, "How Satya Nadella Created a 'Learn-It-All' Culture at Microsoft to Help It Become a $3 Trillion Powerhouse," Business Insider India, July 31, 2024, https://www.businessinsider.in/tech/news/how-satya-nadella-created-a-learn-it-all-culture-at-microsoft-to-help-it-become-a-3-trillion-powerhouse/articleshow/112168591.cms.

8 Lisa Vollmer, "Anne Mulcahy: The Keys to Turnaround at Xerox," Graduate School of Stanford Business, December 1, 2004, https://www.gsb.stanford.edu/insights/anne-mulcahy-keys-turnaround-xerox.

technology into a cohesive offering. Sadoun's foresight in acquiring and integrating companies such as Epsilon and Sapient has not only enhanced Publicis Groupe's capabilities but also redefined the agency's role in a rapidly evolving market. His leadership is marked by a bold vision for the future of marketing, an unwavering commitment to client-centric solutions, and a deep understanding of how technology can be harnessed to drive meaningful human connections.

Implementing Visionary Leadership

To embody the qualities of visionary leadership, leaders should:

- Embrace Strategic Foresight: Develop the ability to anticipate and plan for future industry trends and disruptions.

- Foster Inspirational Leadership: Motivate and inspire teams by communicating a compelling vision and demonstrating commitment to shared goals.

- Demonstrate Resilience and Adaptability: Maintain focus on long-term objectives despite setbacks and remain flexible in response to changing circumstances.

- Make Decisive Decisions: Use a combination of data-driven insights and intuition to make timely and effective decisions.

High-Risk Management

The rapid acceleration of machine intelligence and generative AI has elevated the importance of high-risk management skills for future leaders. As these technologies forge unprecedented opportunities, they also introduce formidable risks, making the ability to effectively

CHAPTER 10

manage and mitigate these risks not merely beneficial but also essential for corporate survival and success.

The sweeping advancements in machine intelligence, exemplified by OpenAI's ChatGPT, promise to revolutionize industries through automating complex tasks, generating nontraditional solutions, and transforming customer interactions. These developments significantly enhance business functions in customer service, content creation, and decision-making. However, they also introduce significant risks, including ethical dilemmas, security vulnerabilities, and potential job displacement as outlined under Principle II, which demand a new caliber of leadership adept at foreseeing and navigating these challenges.

The story of IBM's Watson highlights the dual-edged nature of technological innovation. Initially celebrated for its potential to revolutionize healthcare, it later faced challenges in delivering consistent and reliable medical recommendations.[9] This underscores the critical need for leaders to manage expectations and ensure rigorous testing and validation of AI technologies. It also illustrates the necessity for anticipatory leadership that can foresee potential pitfalls and strategize to mitigate them through proactive risk management and continuous adaptation.

As technology evolves, so, too, must the skills of the workforce. Companies such as Google, Amazon, and Publicis Groupe have led the way with extensive training programs in data science and machine learning. Yet the rapid pace of technological advances can quickly render today's skills obsolete, presenting another layer of risk. Leaders must navigate these uncertainties with a commitment to continuous investment in skill development, fostering a culture of futurism and adaptability within their organizations.

---

9   John Frownfelter, "Why IBM Watson Health Could Never Live Up to the Promises," MedCity News, April 8, 2021, https://medcitynews.com/2021/04/why-ibm-watson-health-could-never-live-up-to-the-promises/.

THE TAO OF LEADERSHIP

Leaders are now challenged to develop a collaborative approach to risk management, a challenging shift for any organization aiming to compete successfully in this dynamic environment. This approach, along with essential leadership qualities, forms the bedrock for navigating and capitalizing on the relentless pace of technological change.

A collaborative approach to risk management involves anticipating potential issues before they arise and implementing strategies to mitigate them effectively. It starts with a comprehensive assessment of the current and future competitive landscape in which an organization operates, identifying potential industry-wide threats from technological disruptions, market changes, economic shifts, workforce displacement, required skills, and regulatory developments. From this assessment, cross-industry collaboration can develop collaborative responses for risk mitigation strategies and continuous monitoring.

The leadership qualities necessary to master risk management include foresight, decisiveness, adaptability, and resilience. Leaders must possess the foresight to predict potential challenges and envision mitigative strategies. They must also be decisive, capable of making tough choices quickly, especially when under pressure. Adaptability is another critical quality, as leaders must pivot their strategies in response to changing circumstances without losing sight of their overarching goals. Lastly, resilience allows leaders to withstand setbacks and continue their pursuit of the organization's objectives.

Several leaders exemplify these qualities and have successfully implemented effective risk management strategies. Mary Barra, CEO of General Motors (GM), is one such leader. Under her leadership, GM has made significant strides in transitioning toward electric vehicles (EVs) and autonomous driving technologies. Recognizing the long-term risks posed by climate change and regulatory shifts, Barra has proactively steered GM to invest heavily in EVs, a move that initially involved high risks given the uncertain market acceptance and significant financial investment.[10] Her foresight, decisiveness, and resilience also enabled her to moderate these strategies as market conditions proved to be less positive than GM had anticipated, maintaining GM as a leader in the automotive industry's transformation.

Lisa Su, CEO of Advanced Micro Devices (AMD), has revived the semiconductor company through her strategic vision and risk management. By focusing on high-performance computing and graphics technologies, she identified niche markets where AMD could compete effectively against much larger rivals. Her approach involved substantial research and development investments, betting on technologies such as Ryzen and EPYC processors, which were risky at the time but ultimately proved highly successful.[11]

Reed Hastings, cofounder and CEO of Netflix, has demonstrated exceptional risk management by transforming the DVD rental business into a streaming giant. Hastings's decision to pivot to streaming well before the market had fully embraced online video consumption involved significant risk. His foresight into consumer behavior and technological trends, along with his decisiveness and adaptability, has made Netflix a dominant force in the entertainment industry.

Rupert Murdoch, the media mogul behind News Corporation, has consistently navigated high-risk environments by diversifying media holdings and embracing digital innovation, even when it involved cannibalizing existing business models. His aggressive strategies in digital media and global expansion have shown a profound understanding of risk management, adapting to new media land-

---

[10] Jonathan Sterling, "General Motors CEO Mary Barra: 'We Believe in an All-Electric-Future,'" Columbia Business School, April 25, 2024, https://business.columbia.edu/insights/climate/general-motors-ceo-mary-barra-we-believe-all-electric-future.

[11] Ian, "Lisa Su's Journey to AMD Leadership," Pressfarm, June 5, 2024, https://press.farm/lisa-sus-journey-to-amd-leadership/.

THE TAO OF LEADERSHIP

scapes, including the sale of Fox's studio and several entertainment assets to focus primarily on live sports and news content.[12]

These leaders illustrate that mastering the components of proactive risk management and embodying the associated leadership qualities can lead organizations not only to navigate but also to thrive amid the challenges posed by the rapid pace of technological advancements. Their success stories provide a blueprint for other leaders aiming to secure their organizations' futures in increasingly volatile technological landscapes.

---

12   News Corp, "K. Rupert Murdoch to Be Appointed Chairman Emeritus of Fox Corporation and News Corp," press release, September 21, 2023, https://newscorp.com/2023/09/21/k-rupert-murdoch-to-be-appointed-chairman-emeritus-of-fox-corporation-and-news-corp/.

CHAPTER 11

Context-Aware AI, Inclusivity, Education, Sustainability, and Globalism

Enhancing Relevance and Acceptance

Developing corporate organizations that are sensitive to the cultural, generational, and social context in which they operate is crucial for their relevance and acceptance. In the AI era, team building, product development, and marketing will involve recognition and respect of cultural norms and values that advance the integration of human activities. For instance, context-aware AI can adapt marketing strategies based on cultural preferences or adjust healthcare recommendations according to regional practices. By understanding and incorporating cultural nuances, AI can provide more relevant and meaningful interactions.

This sensitivity to context highlights the positive role of AI in supporting and augmenting human creativity. Human creativity is deeply rooted in cultural and social experiences, and by recognizing these, AI can help foster more authentic and innovative solutions. For example, in creative industries such as advertising, context-aware AI can generate campaigns that resonate more deeply with diverse audiences, reflecting their unique cultural backgrounds and experiences. This not only improves the effectiveness of such campaigns but also enhances the creative process by introducing new perspectives and insights that might otherwise be overlooked.

Context-aware AI can contribute to societal harmony by reducing cultural misunderstandings and promoting inclusivity. In global businesses and multicultural societies, having AI systems that respect and understand diverse cultural contexts can facilitate better communication and collaboration. This integration supports the idea that AI and human creativity are not just coexisting but are also complementary forces driving innovation and understanding.

Enhancing User Satisfaction and Creativity

Leveraging AI to offer personalized experiences that cater to individual preferences and identities can significantly enhance user satisfaction. Personalization can be based on a variety of factors, including gender, ethnicity, and cultural background, provided that it is done ethically and with respect for privacy. Personalized AI can tailor recommendations for music, movies, and books; create customized learning plans for students; and provide personalized healthcare advice, among other applications.

Personalized AI supports human creativity by providing tools and resources that are specifically tailored to individual needs and preferences. For instance, in education, AI can create personalized learning experiences that cater to a student's strengths and areas for improvement, fostering a more engaging and effective learning environment. This individualized approach can help nurture creativity by allowing students to explore subjects and projects that align with their interests and passions.

In creative fields such as design and art, personalized AI can offer suggestions and tools that reflect the artist's unique style and preferences, enhancing their creative process. By understanding and adapting to individual preferences, AI can serve as a collaborative partner, providing inspiration and new ideas that align with the creator's vision. This collaboration between human creativity and AI-driven personalization demonstrates how technology can enhance and amplify human ingenuity.

Collaborative AI: Ensuring Inclusivity and Broader Perspectives

Engaging with communities and stakeholders from diverse backgrounds in the development of AI systems ensures that these technologies reflect a broader range of experiences and needs. Collaborative AI development involves incorporating feedback and insights from various demographic groups, industries, and cultural contexts, leading to more inclusive and effective AI solutions. This approach not only improves the relevance and usability of AI systems but also fosters trust and acceptance among users.

Collaborative AI development supports human creativity by integrating diverse perspectives and experiences into the design and functionality of AI technologies. This inclusivity ensures that AI tools are designed to address the needs and preferences of a wide range of users, enhancing their creative potential. For example, in healthcare, involving diverse communities in AI development can lead to more culturally sensitive and effective health interventions, ultimately improving patient outcomes and fostering innovation in medical practices.

Collaborative AI development promotes social equity by ensuring that the benefits of AI are accessible to all segments of society. By actively involving underrepresented groups in the development process, AI technologies can be designed to address specific challenges and opportunities faced by these communities. This inclusive approach not only enhances the functionality and relevance of AI systems but also empowers individuals and communities to harness AI for their creative and innovative endeavors.

Future Perspectives in Education

Digital technologies have revolutionized the way knowledge is accessed and disseminated. Online learning platforms, educational apps, and virtual classrooms have made education more accessible, interactive, and personalized. These advancements enable learners to access a wealth of information and resources from anywhere in the world, breaking down geographical and socioeconomic barriers to education.

Drawing insights from Neil Postman's *Teaching as a Subversive Activity,* the need for education to be a dynamic, interactive process that encourages students to question and explore has been emphasized but far from fully realized as educational institutions remain embedded in legacy curricula and institutionalized in outdated organizational models. Postman advocated for a student-centered approach that fosters critical thinking and inquiry, aspects deeply aligned with the capabilities of new media, especially generative AI.

We can hope that this next era of technological innovation will lead to a more "Postmanesque" future.

Generative AI has the potential to revolutionize education by adapting to the learning pace and style of each student, providing personalized resources and support. It can identify strengths and weaknesses, tailoring the educational content accordingly. AI can generate interactive and engaging content, such as simulations, games, or virtual reality experiences, making learning more immersive and effective. Additionally, it can provide instant feedback and assessments, helping students understand their progress and areas for improvement in real time, and assist in research by quickly analyzing vast amounts of data and generating new insights. This personalized approach to education ensures that each student receives the support and resources they need to thrive, regardless of their starting point.

As the education landscape evolves with these technologies, traditional media theories, which often focus on the linear transmission of information and assume a more passive role of the learner, will become less relevant. AI and new media emphasize interactivity and user agency, challenging the one-way communication model of traditional media.

However, integrating generative AI into education is not without its challenges. Educational institutions must navigate the complexities of enabling and empowering the use of generative AI in students' and professors' original writing and reports. There is a fine line between leveraging AI for educational enhancement and ensuring the originality and authenticity of academic work. Institutions must develop robust guidelines and ethical standards to ensure that AI-generated content is used responsibly and does not undermine the integrity of academic work.

The role of AI in research also presents significant opportunities and challenges. While AI can expedite the research process by quickly analyzing vast datasets and generating new insights, there are inherent dangers in the accuracy and reliability of AI-generated content. Researchers must remain vigilant in verifying AI-generated data and ensuring that conclusions drawn from such data are sound and credible. The adoption of past technologies in education, such as the introduction of computers and the internet, faced similar challenges. Initially, there were concerns about the reliability and authenticity of information found online. Over time, these challenges were addressed through the development of digital literacy programs and the establishment of credible online resources. Similarly, the integration of AI into education will require a concerted effort to develop new literacy skills and establish guidelines for responsible use.

Online learning platforms and educational apps are set to become even more accessible, collaborative, adaptive, and immersive. Advances in technology will continue to reduce costs and increase the availability of educational resources, making education more accessible to people around the world and across all socioeconomic boundaries. New tools will facilitate collaboration among students and teachers across geographical boundaries, fostering global learning communities. AI-driven platforms will become more adept at adapting to individual learning needs, providing a truly personalized learning experience. Technologies such as virtual reality and augmented reality will make learning experiences more immersive and engaging, helping to simulate real-world scenarios or visualize complex concepts.

Technology can cater to a wide range of learning styles and needs, including those of students with disabilities and learners in remote areas. With the ease of accessing learning materials and the adaptability of learning paths, individuals are encouraged to engage in lifelong learning, continually acquiring new skills and knowledge throughout their lives.

By bridging the gap between traditional educational models and the innovative potential of new media technologies, we can create a more inclusive, engaging, and effective educational landscape. This evolution not only democratizes access to education but also prepares individuals to thrive in an increasingly complex and dynamic world, where the integration of human empathy, intuition, and ingenuity with advanced technologies will be paramount.

Educational institutions are already adapting to this future by integrating courses on digital media, AI, ethics in technology, and related fields into their curricula. For example, MIT's Media Lab is at the forefront of combining technology with humanistic studies, preparing students to understand the complex interplay between humans and machines. Stanford University's AI Lab is another prime example, focusing on cutting-edge AI research and its applications across various fields, including healthcare, environmental science, and social good. Carnegie Mellon University's Robotics Institute leads in robotics research and AI, exploring both the technical and ethical dimensions of these technologies.

The S. I. Newhouse School of Public Communications at Syracuse University is a distinguished leader in shaping the future of media and communications, earning its place among these forward-thinking institutions. Newhouse is renowned for its innovative approach to education, blending traditional communication disciplines with cutting-edge technology and media studies. The school has been proactive in incorporating courses on digital media, AI, and the ethical implications of emerging technologies into its curriculum. By fostering a deep understanding of the evolving media landscape, Newhouse prepares its students to navigate and lead in an industry

THE TAO OF LEADERSHIP

that is increasingly influenced by technological advancements. The school's emphasis on experiential learning, global perspectives, and interdisciplinary collaboration equips its graduates with the skills and insights necessary to drive meaningful change in the communication field, ensuring they are not only adept at using new technologies but also thoughtful about their broader societal impacts.

The Cathy Hughes School of Communications at Howard University undoubtedly deserves recognition alongside other future-looking educational programs. This school has a long-standing commitment to empowering students through a curriculum that emphasizes both technical proficiency and social justice, making it a leader in producing communication professionals who are not only skilled but also socially conscious. The Cathy Hughes School has been at the forefront of integrating digital media and technology into its programs, ensuring that students are well versed in the latest tools and platforms. Moreover, the school places a strong emphasis on the role of media in shaping public discourse and advancing societal change, particularly within historically underrepresented communities. By combining rigorous academic training with a focus on ethical communication and advocacy, the Cathy Hughes School of Communications equips its graduates to lead and innovate in a media landscape that increasingly values diversity, equity, and inclusion. Its inclusion among these distinguished institutions reflects its pivotal role in preparing future leaders who are capable of navigating and transforming the complex interplay between technology, media, and society.

Institutions such as these are promoting interdisciplinary approaches, combining the study of science, technology, engineering, and mathematics (STEM) with arts and humanities to provide a holistic view of the challenges and opportunities posed by advanced technologies. Leading universities have established research centers focusing on AI, robotics, and digital media. These centers not only contribute to technological advancements but also explore their societal and ethical implications. Think tanks such as the Brookings Institution and the RAND Corporation conduct research on the future of technology, focusing on policy implications, societal impacts, and ethical considerations. These organizations and educational institutions play a crucial role in raising public awareness about the challenges of the digital future and making policy recommendations to governments and international bodies.

To further this progress, more educational institutions must follow these examples. They should invest in interdisciplinary programs that blend technology with humanistic studies, establishing research centers that explore the ethical and societal implications of new technologies. Universities and colleges should also prioritize digital literacy, ensuring that students are equipped with the skills to critically evaluate AI-generated content and use these technologies responsibly.

Corporate leaders also have a vital role to play in supporting education. By investing in educational programs within their companies, they can ensure that their workforce is prepared for the future. This includes offering training in digital literacy, AI, and other emerging technologies. Additionally, corporations can support educational institutions through funding research, sponsoring scholarships, and providing internships and practical experiences for students.

For instance, IBM has been instrumental in supporting educational initiatives by partnering with schools to integrate AI and blockchain into their curricula.[13] Similarly, Intel supports STEM education through its Intel AI for Youth program, which aims to

---

13 Soroush Abbaspour, "Blockchain in Higher Education: Preparing Students for the Future," IBM, November 4, 2019, https://www.ibm.com/blog/blockchain-in-higher-education-preparing-students-for-the-future/.

empower students to learn and apply AI skills.[14] Salesforce also invests heavily in education through its Trailhead platform, which provides free online learning in various tech fields, ensuring that students and professionals alike can upskill and reskill for the future workforce.[15]

It is essential to support a wide spectrum of institutions, including historically black colleges and universities (HBCUs), Hispanic-serving institutions (HSIs), technical schools, community colleges, and military education programs. These institutions play a crucial role in providing diverse and accessible education, and their support is vital for creating a well-prepared and inclusive workforce. For example, the Thurgood Marshall College Fund and the Hispanic Scholarship Fund offer support and scholarships to students at HBCUs and HSIs, ensuring that underrepresented groups have access to quality education and opportunities in emerging fields.

Leading Sustainably in the AI Era

As we embrace the transformative power of AI technologies, it's imperative for corporate leaders to address the environmental impact and carbon footprint associated with these advancements. The energy consumption of AI, particularly in training large models, can be substantial, leading to significant carbon emissions.[16] To ensure that AI

---

14 Livia Martinescu, "Reflections on the Intel AI for Youth Program," Oxford Insights, February 6, 2023, https://oxfordinsights.com/insights/2023-2-6-reflections-on-the-intel-ai-for-youth-program/.

15 Ann Weeby, "What Is Trailhead? All About Salesforce's Free Online Learning Platform," Salesforce Insider, February 2, 2024, https://salesforceinsider.com/what-is-trailhead-all-about-salesforces-free-online-learning-platform/.

16 Renée Cho, "AI's Growing Carbon Footprint," State of the Planet, Columbia Climate School, June 9, 2023, https://news.climate.columbia.edu/2023/06/09/ais-growing-carbon-footprint/.

CHAPTER 11

development and deployment align with sustainability goals, leaders must adopt a proactive and responsible approach.

The training of AI models, especially deep learning models, requires vast amounts of computational power. Data centers, which house the necessary hardware for AI training and operations, consume a tremendous amount of energy. This energy use not only increases operational costs but also contributes to the carbon footprint of organizations.[17] A study by the University of Massachusetts Amherst found that training a single AI model can emit as much carbon as five cars over their lifetimes.[18] Such statistics highlight the urgent need for sustainable practices in AI development.

Corporate leaders can implement several strategies to mitigate the environmental impact of AI. Transitioning to energy-efficient data centers can significantly reduce the carbon footprint of AI operations. Companies such as Google and Microsoft have invested in renewable energy sources to power their data centers. Google has achieved 100 percent renewable energy for its global operations, setting a benchmark for the industry.

Developing more efficient algorithms that require less computational power for training and inference can also help. Researchers are exploring techniques such as model compression and federated learning to reduce energy consumption. Additionally, investing in carbon offset projects can neutralize the emissions generated by AI operations. Amazon, for example, has committed to net-zero carbon

---

17   Kim Martineau, "Shrinking Deep Learning's Carbon Footprint," MIT News, August 7, 2020, https://news.mit.edu/2020/shrinking-deep-learning-carbon-footprint-0807.

18   Stefen Joshua Rasay et al., "AI's Large Carbon Footprint Poses Risks for Big Tech," S&P Global, October 17, 2019, https://www.spglobal.com/marketintelligence/en/news-insights/latest-news-headlines/ai-s-large-carbon-footprint-poses-risks-for-big-tech-54710079.

by 2040 and is investing in various sustainability initiatives, including reforestation and renewable energy projects.

Ensuring that the hardware used for AI, from servers to chips, is produced and disposed of sustainably is crucial. Companies should work with suppliers who adhere to environmental standards and promote the recycling and reuse of electronic components. Supporting research into green AI technologies and methodologies is essential, and collaborative efforts between academia and industry can lead to breakthroughs in sustainable AI.

Several companies and leaders are at the forefront of integrating sustainability into AI development. Google is also working on AI models that require less computational power and investing in carbon offset projects. Microsoft's AI for Earth initiative aims to leverage AI to solve environmental challenges. The company has committed to being carbon negative by 2030 and is investing in renewable energy, reforestation, and sustainable AI research. IBM's Green Horizon project uses AI to improve air quality management and forecast pollution levels. IBM is also focused on reducing the energy consumption of its AI systems and data centers. Salesforce has integrated sustainability into its core business strategy, with initiatives such as its Sustainability Cloud, which helps organizations track and reduce their carbon footprint. Salesforce integrated sustainability into its core business strategy through its Net Zero Cloud (formerly known as Sustainability Cloud), which helps organizations track and reduce their carbon emissions. One year ahead of its goal, the company achieved 100 percent renewable energy across its global operations in 2021 and has committed to becoming a net-zero emissions company across its entire value chain. The company is investing its carbon removal projects and reforestation efforts aimed at planting 100 million trees by 2030.[19]

Project Dandelion exemplifies the power and influence of women in driving change. This initiative aims to harness AI for global social impact in advancing environmental activism. By focusing on empowering and leveraging the collective strength and leadership of women and applying marketing impact to achieve sustainability goals across the globe, Project Dandelion is fostering significant advancements in addressing environmental risks to humanity. This collective approach can amplify the impact and drive systemic change.

By adopting energy-efficient practices, investing in green technologies, and leading by example, corporations can harness the power of AI to drive innovation and progress while safeguarding our planet for future generations.

Global Perspectives on AI

As AI continues to evolve, its adoption and implementation vary significantly across different regions and cultures. These variations are influenced by unique societal norms, economic conditions, regulatory environments, and cultural attitudes toward technology. Understanding these diverse approaches is crucial for leaders aiming to harness the global potential of AI while addressing its broader implications.

In North America, particularly the United States, AI development is driven by a robust private sector and significant investments from tech giants such as Google, Microsoft, and Amazon. The United States emphasizes innovation and entrepreneurialism, fostering a competitive environment where AI technologies rapidly advance.

---

19   Max Scher, "Forging the Path to 100% Renewable Energy, Together," Salesforce, January 17, 2019, https://www.salesforce.com/blog/path-to-renewable-energy/.

However, this rapid pace often outstrips regulatory frameworks, leading to concerns about privacy, security, and ethical considerations. Canadian AI strategy, on the other hand, balances innovation with ethical AI principles. Canada has been a pioneer in AI ethics, with initiatives such as the Montréal Declaration on Responsible AI, which advocates for the responsible use of AI technologies.

Europe takes a more cautious and regulatory approach to AI adoption. The European Union has introduced the General Data Protection Regulation, setting stringent standards for data privacy and security. The EU's proposed AI regulations aim to ensure that AI systems are transparent, traceable, and accountable. This focus on ethical AI seeks to protect citizens' rights while fostering innovation. Countries such as Germany and France are investing heavily in AI research and development, aiming to become leaders in ethical AI and sustainable technology.

In Asia, China is a dominant force in AI development, driven by massive state investment and a strategic focus on becoming the global leader in AI by 2030. The Chinese government's AI strategy prioritizes sectors such as surveillance, healthcare, and transportation, leveraging AI to boost economic growth and enhance public services. However, China's approach raises global concerns about surveillance, privacy, and the use of AI for state control. Japan, by contrast, emphasizes AI's role in addressing societal challenges such as an aging population. Japan's Society 5.0 initiative envisions a supersmart society where AI and robotics enhance quality of life, promoting harmony between technological advancement and human welfare.

India is emerging as a significant player in the AI landscape, with a focus on using AI to address socioeconomic challenges. The Indian government's National Strategy for Artificial Intelligence aims to leverage AI for inclusive growth, emphasizing healthcare, agriculture, education, and smart cities. India's approach highlights the potential of AI to drive development in emerging economies and improve living standards for millions.

The Middle East, particularly the United Arab Emirates and Saudi Arabia, is investing heavily in AI as part of broader economic diversification strategies. The UAE's AI strategy aims to position the country as a global hub for AI innovation, integrating AI across sectors such as transportation, healthcare, and public services. Saudi Arabia's Vision 2030 outlines the country's plans to harness AI for economic transformation and societal advancement.

Africa presents a unique perspective on AI adoption, characterized by grassroots innovation and a focus on solving local challenges. Countries such as Kenya and Nigeria are leveraging AI to improve agriculture, healthcare, and financial services, addressing issues such as food security and access to quality healthcare. African AI initiatives often emphasize community-driven solutions and the potential of AI to foster sustainable development.

The global implications of AI are profound, necessitating international collaboration and regulation. As AI technologies cross borders, there is an urgent need for harmonized standards and ethical guidelines to ensure that AI benefits humanity. International bodies such as the United Nations and the Organisation for Economic Co-operation and Development are working to develop global frameworks for AI governance, emphasizing transparency, accountability, and inclusivity.

Collaboration among nations is essential to address the challenges and opportunities presented by AI. Shared research initiatives, cross-border data sharing, and cooperative regulatory efforts can help mitigate risks while maximizing the benefits of AI. Leaders must advocate for policies that promote the responsible development and deployment of AI, ensuring that technological advancements are aligned with global ethical standards and human values.

The cultural context in which AI is developed and implemented plays a critical role in shaping its impact. Understanding and respecting cultural differences can enhance the effectiveness of AI solutions and foster global trust in AI technologies. Leaders must cultivate a global mindset, recognizing that AI's potential to drive progress is amplified when diverse perspectives and expertise are integrated.

The global perspectives on AI highlight the diverse approaches and priorities that different regions bring to AI development and adoption. As we navigate this transformative era, it is imperative for leaders to embrace a global vision, promoting ethical AI that respects cultural differences and enhances human well-being worldwide. The efforts of initiatives such as Project Dandelion and the leadership of women in AI further underscore the importance of inclusivity and collaboration in shaping a sustainable and equitable future.

PRINCIPLE V

Integrity—Organizational Consolidation Across Capabilities

CHAPTER 12

The FusionFlow Process

As technological advancements accelerate, consolidation of teams across varying skill sets, capabilities, and responsibilities is critical not only for enhancing efficiency and communication within organizations but also for integrating AI and machine learning into every facet of business operations.

Consolidation allows organizations to break down traditional silos that hinder collaboration and stifle creativity. Integrated and versatile teams ensure that data and insights are shared seamlessly, enabling quicker, more informed decision-making processes, especially in environments that thrive on the free flow of vast, integrated datasets.

The FusionFlow Process is a strategy developed as part of the Myers Blueprint for Leadership in the AI age. While the term itself is new, the practice has been implicitly applied by forward-thinking companies for decades, demonstrating its effectiveness in fostering sustainable growth and adaptability.

FusionFlow conveys the idea of merging or integrating different elements (fusion) while maintaining a smooth, continuous process (flow). This suggests a dynamic and harmonious approach to consolidation, breaking down silos and ensuring stability in a way that's easy to understand and implement within an organization.

The FusionFlow Process is about creating a balanced approach within organizations that accelerates change while ensuring stability. This strategy involves integrating innovative practices that disrupt traditional business models without compromising the core operations that ensure the company's day-to-day stability. It's about harmonizing the old with the new, allowing companies to explore new frontiers without losing their foundational strengths.

Several leading companies across industries have exemplified the FusionFlow Process approach.

Sundar Pichai at Google has exemplified this transformative approach. Under Pichai's leadership, Google underwent significant cultural and organizational changes that dismantled existing silos and fostered a collaborative, data-driven culture. This shift was instrumental in enabling Google to fully utilize AI and machine learning across its various products and services, solidifying its position as a leader in cloud computing and AI solutions.

In the communications industry, Tom Bartlett, the former CEO of American Tower Corporation—a leading independent owner, operator, and developer of multitenant communications real estate—also championed such integrative efforts. Recognizing the potential of AI to transform infrastructure management, Bartlett oversaw initiatives aimed at breaking down operational silos within the company. This enabled American Tower to enhance efficiency in site management and tenant communications and better leverage data analytics for decision-making.[20] Under current CEO Steven Vondran, these changes remain

---

20   Paul Lipscombe, "American Tower CEO and President Tom Bartlett to Retire," DCD, October 27, 2023, https://www.datacenterdynamics.com/en/news/american-tower-ceo-and-president-tom-bartlett-to-retire/.

pivotal as the company expands its global footprint and adapts to the rapidly evolving demands of wireless technology services.

Reed Hastings's leadership transformed Netflix from a DVD rental service to a streaming powerhouse by effectively integrating AI into its core operations. This strategic move allowed Netflix to personalize content recommendations and optimize streaming services, fundamentally changing the media and entertainment landscape.[21] Hastings's success in consolidating organizational functions and fostering a unified, technology-driven culture was key to this achievement.

On a smaller scale, Lemonade, an insurance start-up, demonstrates how effective consolidation can disrupt traditional industries. Under CEO Daniel Schreiber, Lemonade adopted a flat organizational structure that emphasized agility and rapid innovation.[22] This approach enabled the seamless integration of AI into their operations, facilitating instant claims processing and personalized service offerings.[23, 24]

Apple has consistently managed to introduce groundbreaking technologies and products while maintaining robust operational processes that ensure smooth customer experiences and steady revenue streams. Their approach to seamlessly integrating innovative hardware and software advancements with strong supply chain management exemplifies the FusionFlow Process.

---

21   Peter Westberg, "Reed Hastings: The Architect of Netflix's Rise," Quartr, April 11, 2024, https://quartr.com/insights/business-philosophy/reed-hastings-the-architect-of-netflixs-rise.

22   "Interview: Daniel Schreiber, Co-Founder and CEO of Lemonade," FinTech Futures, June 5, 2017, https://www.fintechfutures.com/2017/06/interview-daniel-schreiber-co-founder-and-ceo-of-lemonade/.

23   Laura Heely et al., "Award Winner: Delighting Insurance Customers with AI and Behavioural Economics," Case Centre, February 27, 2023, https://www.thecasecentre.org/caseSpotlight/2023/Lemonade.

24   Shai Wininger, "The Secret Behind Lemonade's Instant Insurance," Lemonade, n.d., https://www.lemonade.com/blog/secret-behind-lemonades-instant-insurance/.

Known for its Toyota Production System, Toyota has applied principles of continuous improvement and innovation while ensuring rigorous standards of quality and efficiency. Toyota's method balances the integration of new technologies and processes in manufacturing, such as robotics and AI, with a deep-rooted corporate culture that prioritizes reliability and worker empowerment.[25, 26]

IBM, as a company that transitioned from hardware to a leader in cloud computing and AI, has continually evolved its business model. The company maintains stability through its global services and consulting divisions while aggressively investing in new technologies such as quantum computing and blockchain.[27]

An advertising agency holding company that has been recognized for effectively consolidating its capabilities, organizations, and resources is **Publicis Groupe.** Under the leadership of CEO Arthur Sadoun, Publicis Groupe has successfully implemented a strategy known as the "Power of One," which aims to break down operational silos and create a more integrated, collaborative, and client-focused organization.[28]

---

25 Jamie P. Bonini et al., "Building an Engaging Toyota Production System Culture to Drive Winning Performance for Our Patients, Caregivers, Hospitals, and Communities," in *Patient Safety and Quality Improvement in Healthcare*, eds. Rahul K. Shah and Sandip A. Godambe (Springer, 2020).

26 John Shook, "How Toyota Built Its Culture Around the World," Lean Enterprise Institute, September 6, 2021, https://www.lean.org/the-lean-post/articles/how-toyota-built-its-people-centric-culture/.

27 "EY and IBM Expand Strategic Alliance into Quantum Computing," IBM, April 13, 2023, https://newsroom.ibm.com/2023-04-13-EY-and-IBM-expand-strategic-alliance-into-quantum-computing.

28 "The Supervisory Board of Publicis Groupe Renews the Mandate of Arthur Sadoun as Chairman and CEO of Publicis Groupe," Publicis Groupe, September 14, 2022, https://www.publicisgroupe.com/en/news/press-releases/the-supervisory-board-of-publicis-groupe-renews-the-mandate-of-arthur-sadoun-as-chairman-and-ceo-of-publicis-groupe.

CHAPTER 12

Publicis Groupe's consolidation efforts involved restructuring its operations to unify its various agencies and specialized services under a single umbrella. This integration allowed the company to leverage its vast resources across data, creativity, media, and technology more efficiently. By aligning its capabilities, Publicis Groupe improved internal collaboration, reduced redundancies, and enhanced its ability to deliver comprehensive, end-to-end solutions for clients.[29]

A key component of this strategy was the acquisition and integration of companies such as Epsilon, a data-driven marketing powerhouse, and Sapient, a leader in digital transformation. These acquisitions were seamlessly incorporated into Publicis Groupe's operations, enabling the company to offer a unified platform that combines deep consumer insights, creative expertise, and advanced technology solutions.[30]

The result of these efforts has been increased operational efficiency, improved business results, and a stronger competitive position in the market.[31] Publicis Groupe's ability to effectively consolidate its capabilities and break down silos has set a benchmark in the industry for how large holding companies can adapt to the demands of a rapidly changing advertising landscape.

Dentsu has recently taken steps to consolidate its operations, including the full integration of Merkle, its data and analytics division, into its broader organizational structure. In 2023 Dentsu announced

---

29 Campaign Staff, "Publicis Groupe's Restructure Shows Much Needed Change, but Hard Facts Need More Clarity," Campaign Middle East, December 20, 2015, https://campaignme.com/publicis-groupes-restructure-shows-much-needed-change-but-hard-facts-need-more-clarity/.

30 "Publicis Groupe Finalizes the Acquisition of Epsilon," Epsilon, July 2, 2019, https://www.epsilon.com/us/about-us/pressroom/publicis-groupe-finalizes-the-acquisition-of-epsilon.

31 "Publicis Groupe: 2019 Full Year Results," Business Wire, February 6, 2020, https://www.businesswire.com/news/home/20200205005907/en/Publicis-Groupe-2019-Full-Year-Results.

THE TAO OF LEADERSHIP

that it would unify its operations under a single brand identity—**dentsu**—bringing together its global network of agencies, including Merkle, to operate under a more cohesive and streamlined structure.[32]

This latest consolidation effort is part of dentsu's ongoing strategy to simplify its organizational structure, enhance collaboration, and leverage its data and technology capabilities across the entire network. By fully integrating Merkle, which had previously operated as a semi-autonomous entity within dentsu, the company has strengthened its ability to deliver data-driven, customer-centric marketing solutions at scale.

The integration of Merkle into the dentsu brand underscores the company's commitment to breaking down silos and creating a unified approach to serving clients. This move allows dentsu to more effectively combine its expertise in creative, media, and customer experience management, offering clients a seamless and comprehensive service.[33] The rebranding effort reflects dentsu's evolution into a more agile and integrated organization, poised to meet the challenges of a rapidly changing marketing landscape.

The success of this consolidation is evidenced by dentsu's improved ability to deliver holistic marketing solutions that are deeply rooted in data and technology, as well as by its strengthened market position as a leading global advertising and marketing communications network.[34]

---

32 "Dentsu Releases 'dentsu Integrated Report 2023,'" dentsu, July 31, 2023, https://www.group.dentsu.com/en/news/release/001003.html.

33 "Dentsu Strengthens Customer Transformation & Technology Capabilities with the Acquisition of Aware Services," Merkle, December 14, 2022, https://www.merkle.com/en/merkle-now/press-releases/2022/dentsu-strengthens-customer-transformation-technology-capabilities.html.

34 "Innovating to Impact: Dentsu Unveils New Global Brand Proposition Drawing upon 120 Year Heritage," GlobeNewswire, May 16, 2024, https://www.globenewswire.com/news-release/2024/05/16/2883594/0/en/Innovating-to-Impact-Dentsu-Unveils-New-Global-Brand-Proposition-Drawing-Upon-120-Year-Heritage.html.

These examples highlight the indispensable role of organizational consolidation in navigating the complexities of technological transformation. Leaders must be adept at reconfiguring corporate structures, ecosystems, and employee roles to stay aligned with technological progress. With rapid technological advancements continually disrupting established industries, organizations face the dual challenge of driving change while maintaining a stable equilibrium.

Integrating the FusionFlow Process into legacy business models involves a multistep approach to ensure smooth transition and full integration across divisions.

1. Strategic Assessment: Begin with a thorough analysis of current operations, identifying core areas where creativity is actively employed and where stability is most dependent on legacy models and leadership. This assessment should also include business areas that are most positively impacted by emerging AI technologies and those that are most at risk.

2. Incremental Integration: Instead of overhauling entire systems, introduce AI and machine intelligence tactically where they are most opportune and least disruptive. This allows the organization to test and learn from change without overwhelming existing processes. For instance, pilot projects or limited-scope deployments can provide insights and data to guide further integration.

3. Cross-Departmental Collaboration: Establish cross-functional teams that bring together expertise from various departments for deployment of specific projects. Cross-functional teams are crucial for ensuring that advanced technology is effectively integrated into all areas of the business without disrupting core functions.

4. Continuous Learning and Adaptation: Foster a culture of continuous learning and flexibility within the organization. Encourage feedback and adaptation as new technologies and processes are integrated, using insights gained to refine and improve the approach.

5. Leadership and Vision Alignment: Ensure that all levels of leadership are aligned with the vision of incrementally introducing technological advances across the organization on a project-by-project and employee-by-employee basis. This alignment is crucial for maintaining organizational coherence and commitment as changes are implemented. Team members who actively resist this process can be systematically removed from authority and responsibility and ultimately replaced.

By adopting the FusionFlow Process, companies can navigate the complexities of modern technological landscapes more effectively, supporting sustained innovation while preserving organizational stability.

Adopting a corporate commitment focused on consolidation of organizational models across divisions and capabilities is essential in a technology-driven world. Leaders must champion this consolidation to create agile, collaborative environments capable of harnessing the power of AI and machine learning. By fostering a culture of creativity and ensuring continuous education and upskilling, leaders can equip their organizations to navigate the challenges and opportunities presented by technological advancements, ensuring long-term success and sustainability.

CHAPTER 12

Integrating Supply Chains: Lessons from Walmart's Success

With AI and machine intelligence reshaping whole industries, it's crucial for leaders to understand that partners and vendors in a supply chain must adapt to the same policies and principles as their own teams and employees. This alignment is essential for creating a seamless, efficient, and responsive supply chain that can meet the demands of today's market.

Walmart's supply chain management program (eSellerHub) is renowned for its efficiency, innovation, and ability to predict demand and manage inventory effectively. This success can be attributed to several key strategies that have enabled Walmart to collapse silos and foster collaboration both internally and with its supply chain partners. At the core of Walmart's approach is the integration of advanced AI-driven technologies, which facilitate real-time data sharing and analysis across the entire supply chain. This integration ensures that all stakeholders, from suppliers to end customers, are operating with the same up-to-date information, allowing for better decision-making and more accurate demand forecasting.[35, 36]

One of the primary reasons for Walmart's success is its emphasis on creating a cohesive system where silos are dismantled. Internally, Walmart has established cross-functional teams that work collaboratively to achieve common goals. This approach has been extended to include supply chain partners, ensuring that vendors and suppliers

---

35 "Key to Having a Successful Walmart Supply Chain Management," eSellerHub, January 19, 2021, https://www.esellerhub.com/blog/key-to-having-a-successful-walmart-supply-chain-management/.

36 Hugo Britt, "The Secret to Managing the Walmart Supply Chain, One of the Most Effective Supply Chains in the World," Thomas Publishing Company, July 31, 2020, https://www.thomasnet.com/insights/walmart-supply-chain/.

are not encumbered by outdated, siloed, and unconsolidated organizations.[37] By fostering a culture of collaboration and transparency, Walmart has been able to build a supply chain that is agile, resilient, and highly responsive to market changes.

Leaders in the age of AI and machine intelligence must take a proactive role in reviewing and rebuilding their supply chains to align with companies, partners, and leaders who share a similar vision and perspective of the future. This involves adopting models such as the FusionFlow Process, which balance the need for sustained innovation with the preservation of organizational integrity and stability. By aligning supply chain partners with this model, companies can navigate the complexities of modern technological landscapes more effectively.

In supply chain management, disruptions can have significant ripple effects. Leaders who adopt the FusionFlow model can ensure that their supply chains remain robust and adaptable, capable of leveraging AI and other advanced technologies to enhance efficiency and responsiveness.

Beyond internal operations, effective leaders recognize the importance of extending AI integration to supply chain vendors and suppliers. This holistic approach ensures a streamlined flow of data across the value chain, enhancing overall efficiency and responsiveness. For example, Walmart's AI-driven supply chain management strategies have significantly improved its ability to predict demand and manage inventory more effectively. By utilizing AI to analyze sales data, weather patterns, and other factors, Walmart can anticipate customer needs and adjust inventory levels accordingly, reducing

---

[37] Tom Wells, "An Inside Look at Walmart's World-Class Supply Chain Strategy," Marketing Scoop, August 23, 2024, https://www.marketingscoop.com/consumer/walmart-supply-chain-strategy/.

CHAPTER 12

waste and ensuring that products are available when and where they are needed.[38]

The case for leaders to align their supply chains with like-minded partners is further strengthened by the benefits of a unified approach to technology and data management. When all parts of the supply chain are operating with the same principles and technologies, the result is a more cohesive and efficient system. This alignment reduces the risk of miscommunication, delays, and other issues that can arise from working with partners who operate under different paradigms.

Breaking Down Silos

The imperative to dismantle organizational silos is critical as businesses evolve in the age of machine intelligence. Silos, traditionally created to manage complexity and scale within decentralized companies, are now proving to be major obstacles in an era characterized by rapid technological advancements. These isolated structures restrict the free flow of information, hinder collaboration, and stifle innovation, leading to inefficiency and a lack of adaptability that can stagnate a company's growth.[39]

Peter Drucker's *Concept of the Corporation* illustrates how post–World War II manufacturing companies employed decentralized, siloed structures effectively during the mid-twentieth century. However, this model has become counterproductive in today's AI-infused business

---

38 "Walmart: AI Helps Create 'Ready for Anything' Supply Chains," PYMNTS, July 3, 2024, https://www.pymnts.com/supply-chain/2024/walmart-ai-helps-create-ready-for-anything-supply-chains/.

39 Andrew Spanyi, "Succeed with Digital Transformation by Breaking Down Silos," Cognitive World, January 10, 2021, https://cognitiveworld.com/articles/2021/1/6/succeed-with-digital-transformation-by-breaking-down-silos.

environment. Breaking down silos is crucial for integrating machine intelligence and fostering an agile, innovative workplace.

Adobe's transition from traditional software sales to a cloud-based subscription model required breaking down significant internal silos, particularly between their creative software divisions and their marketing and document management teams. This shift was crucial for integrating AI and machine learning into their products effectively, such as their Creative Cloud suite and the Adobe Document Cloud.[40]

By breaking down these silos, Adobe not only improved internal efficiencies but also succeeded in delivering a more cohesive and integrated user experience. They were able to utilize AI to offer innovative features such as image recognition in Adobe Stock and enhanced document intelligence services in Adobe Acrobat, which have significantly contributed to their market dominance in the creative and document management sectors.

This reorganization also allowed for a more agile response to customer needs, quicker iteration on product development, and a more unified approach to data and analytics, which are critical in the AI era. Adobe's successful transformation and adoption of AI technologies demonstrate the effective execution of silo breakdown, leading to tangible business outcomes.

Adobe's strategic decision to break down organizational silos has been a cornerstone of its ability to adapt to and capitalize on technological disruption, ultimately driving significant growth. This transformation began in earnest when Adobe shifted from selling boxed software to a cloud-based subscription model under the Adobe Creative Cloud. This move changed not only how its products were

---

40   Mark Garret, "Reborn in the Cloud," interview by McKinsey, McKinsey Digital, July 1, 2015, https://www.mckinsey.com/capabilities/mckinsey-digital/our-insights/reborn-in-the-cloud.

delivered but also how they were developed and improved, necessitating a more integrated approach within the company.[41]

Adobe's transition required the integration of various functional teams, including software developers, marketing personnel, customer support, and product management. This integration facilitated a seamless flow of information across divisions that were previously siloed, enabling Adobe to synchronize its product development cycles with marketing and customer feedback loops more effectively. For example, insights from real-time data analytics could be quickly looped back to product teams, allowing for rapid iteration of features based on actual user engagement metrics.

The breakdown of silos at Adobe has also been pivotal in its adoption and integration of AI and machine learning technologies across its product suite. Adobe Sensei, the company's AI and machine-learning platform, is a prime example. Sensei powers intelligent features across all Adobe products, enhancing capabilities in imaging, analytics, and automation. For instance, in Adobe Photoshop, Sensei enables features such as content-aware fill and enhanced filter options, which automate complex processes to enhance productivity for users.

The integration of AI into Adobe's cloud services has improved customer experiences by providing personalized tools and content. In Adobe Experience Cloud, AI analyzes customer data to deliver personalized marketing content across various digital platforms. This capability is bolstered by a unified internal structure that supports the rapid development and deployment of AI-driven innovations, ensuring that Adobe stays ahead of customer expectations and market trends.

---

41 Joe Devanesan, "Adobe's Decade of Digital Transformation to the Cloud," TechHQ, April 16, 2021, https://techhq.com/2021/04/adobe-decade-of-digital-transformation-to-the-cloud/.

THE TAO OF LEADERSHIP

The dissolution of internal barriers has also allowed Adobe to be more agile in its product development. By fostering a culture of collaboration, Adobe teams are better equipped to share insights and innovations, speeding the development process and reducing redundancies.[42] This agility was crucial during the rapid shift to digital and remote workflows induced by the COVID-19 pandemic, where Adobe quickly introduced new features and tools to help users adapt to changing work environments.[43]

Economically, Adobe's strategy has driven substantial growth. The shift to a subscription model, supported by continuous product enhancements enabled by AI, has led to a steady increase in recurring revenue. This model has not only stabilized Adobe's revenue streams but has also provided the financial predictability necessary to invest further in AI and other emerging technologies.

Adobe's proactive dismantling of silos has equipped it to respond effectively to technological disruptions, particularly in AI and cloud computing. This strategic realignment has not only enhanced product innovation and customer satisfaction but has also placed Adobe on a path of sustained growth and industry leadership.[44]

---

42   Mark Garret, "Reborn in the Cloud."

43   Jake Brereton, "How They Launched It: Adobe's Cloud Transformation," Launch Notes, January 26, 2021, https://www.launchnotes.com/blog/how-they-launched-it-adobes-cloud-transformation.

44   "Adobe's Promising Future: Strategic Positioning and Growth Prospects in AI and Digital Design," Business Insider, November 16, 2023, https://markets.businessinsider.com/news/stocks/adobe-s-promising-future-strategic-positioning-and-growth-prospects-in-ai-and-digital-design-1032829069.

CHAPTER 12

Squads, Tribes, and Guilds

The squads, tribes, and guilds organization model, initially popularized by Spotify, is designed to promote agility and cross-functional collaboration within a company.[45]

Squads are small, autonomous teams responsible for specific aspects of the product or service. Each squad operates like a mini-start-up, with all the skills needed to design, develop, test, and release their part of the product.[46] Tribes are collections of squads that work in related areas and share a common mission. A tribe can consist of multiple squads that collaborate and share knowledge to ensure alignment and integration of their work.[47]

Guilds are communities of practice that cut across the organization, bringing together individuals with similar roles or expertise. Guilds focus on sharing best practices, fostering learning, and improving specific skills, regardless of the tribes or squads to which members belong to.[48]

This model fosters innovation and accelerates problem-solving by leveraging diverse skills and perspectives within a centralized, flexible, project-focused structure. Despite Spotify's initial success, the intention to increase corporate agility ultimately failed for several reasons. The rapid growth of the company led to scalability issues. As Spotify expanded, the loosely defined roles and responsibilities within squads

---

45 Mark Cruth, "Discover the Spotify Model," Atlassian, n.d., https://www.atlassian.com/agile/agile-at-scale/spotify.

46 Gennaro Cuofano, "Spotify Squad Model," FourWeekMBA, April 19, 2024, https://fourweekmba.com/spotify-squad-model/.

47 Cuofano, "Spotify Squad Model."

48 Cuofano, "Spotify Squad Model."

and tribes caused confusion and inefficiencies.[49] The lack of clear accountability made it challenging to manage dependencies among teams, resulting in delays and misaligned priorities. Additionally, the cultural shift required to sustain such a fluid organizational structure was underestimated.[50] While the model promoted autonomy and innovation, it also demanded a high level of discipline, communication, and alignment among teams. Many employees struggled to adapt to the new ways of working, leading to inconsistencies in how the model was implemented across the organization. The lack of standardized processes and tools created friction and hindered collaboration.[51]

Despite these challenges, the squads, tribes, and guilds model is a viable leadership strategy for organizations seeking greater agility. The principles behind squads, tribes, and guilds—such as cross-functional collaboration, autonomy, and flexibility—are still relevant and can drive significant benefits if implemented correctly.

To implement the process and avoid failures, it is crucial to define clear roles and responsibilities. Clearly outlining the roles and responsibilities within squads, tribes, and guilds ensures that each team member understands their specific duties and how they contribute to the overall goals. Establishing clear leadership structures within each tribe and squad helps maintain accountability and streamline decision-making processes.

Fostering a strong organizational culture is also essential. Promoting a culture of trust, transparency, and continuous learning

---

49 Mark Littlewood, "Lessons from a Critique | Spotify's Failed Squad Model," Business of Software, June 18, 2020, https://businessofsoftware.org/2020/06/lessons-critique-spotifys-failed-squad-model/.

50 Littlewood, "Lessons from a Critique."

51 Arif Harbott, "Why Copying Spotify's Squads and Tribes Model Probably Won't Work for You," August 3, 2020, https://harbott.com/why-squads-and-tribes-probably-wont-work/.

encourages open communication and regular feedback to ensure alignment and address issues promptly. Investing in training programs to help employees adapt to the new organizational structure and develop the necessary skills for effective cross-functional collaboration can further support this effort.

Implementing standardized processes and tools across squads and tribes ensures consistency and facilitates collaboration. This includes project management tools, communication platforms, and reporting systems. Regularly reviewing and updating these processes keeps them aligned with the evolving needs of the organization.

Managing dependencies effectively is another critical factor. Identifying and managing dependencies between squads and tribes prevents bottlenecks and misalignments. Using visual management tools such as dependency maps and Kanban boards to track and coordinate interteam dependencies can be highly effective. Establishing regular synchronization meetings, such as tribe lead meetings and cross-squad coordination sessions, ensures alignment and addresses interteam challenges. While promoting autonomy within squads, it is essential to ensure alignment with the organization's overall strategy and goals. Using objectives and key results to set clear objectives and measure progress across squads and tribes helps achieve this balance. Implementing a governance framework provides guidance and oversight without stifling innovation and autonomy.

Fostering a culture of continuous improvement encourages teams to regularly reflect on their processes and outcomes. Using retrospectives and postmortems to identify areas for improvement and implement changes supports this culture. Celebrating successes and learning from failures creates a learning organization that continuously evolves and adapts.

To further enhance the agility and stability of the organization, implementing the FusionFlow Process can provide a balanced strategy for maintaining stability while driving innovation. This approach focuses on achieving equilibrium by acting as a catalyst for change and ensuring that innovation does not disrupt the overall organizational stability.

Key steps for implementing the FusionFlow Process include establishing a vision for equilibrium, creating a change management framework, promoting leadership and accountability, implementing feedback loops, encouraging collaborative innovation, aligning goals and incentives, and continuously evaluating and adapting. By integrating the FusionFlow Process with the Spotify "squads, tribes, and guilds" model, organizations can achieve a balanced strategy that promotes innovation while maintaining the stability required for short-term success. This holistic approach ensures that teams are empowered to innovate while staying aligned with the organization's overall vision and objectives.

Flat Organizational Structures

On a smaller scale, Basecamp presents another successful case of silo removal through its focus on a flat organizational structure for project management. This approach prioritizes open communication and collaboration, with cofounders Jason Fried and David Heinemeier Hansson advocating for transparency and autonomy. This project management process not only boosts employee engagement but also cultivates an environment where innovation is driven by the collective intelligence of the workforce.

CHAPTER 12

Dynamic Team Structures and Project-Based Work

The traditional hierarchical model of static roles and responsibilities is obsolete. Instead, successful organizations are pivoting toward dynamic team structures and project-based work. This shift is not merely a trend; it's also a fundamental transformation necessary to foster adaptability, innovation, and resilience in the face of rapidly changing market realities.

Organizations have traditionally relied on well-defined roles and responsibilities to maintain order and efficiency. This model often led to siloed departments and a lack of cross-functional collaboration. Employees become pigeonholed into specific tasks, limiting their ability to contribute beyond their designated roles. This rigidity can be particularly detrimental in industries where rapid innovation and adaptation are essential for survival.

The rigidity of fixed roles within traditional organizational structures can stifle innovation and slow response times to market changes. In contrast, dynamic team structures allow for fluid roles and responsibilities, enabling organizations to pivot quickly and efficiently. This approach is critical in an environment where market conditions, consumer preferences, and technological advancements are in constant flux.

Dynamic team structures address these realities by promoting a more flexible approach to work. Instead of static roles, employees are assigned to projects based on their skills, interests, and the current needs of the organization. This fluidity allows for the rapid formation and dissolution of teams, ensuring that the right mix of talents is always available to tackle emerging challenges and opportunities.

One of the primary benefits of dynamic team structures is enhanced adaptability. In a static structure, employees are often confined to narrowly defined roles, which can hinder the organization's ability to adapt to new challenges or opportunities. Dynamic teams, however, can be reconfigured as needed, bringing together individuals with diverse skill sets to address complex projects or problems. This flexibility not only enhances the organization's ability to respond to change but also fosters a culture of continuous learning and innovation.

Consider a scenario where a sudden market shift demands a new product feature targeting a new consumer base. In a traditional setup, the process of reallocating resources, reassigning tasks, reassigning marketing teams, and navigating bureaucratic hurdles can be time-consuming and disruptive. In contrast, a dynamic team structure allows for the swift assembly of a cross-functional team, composed of members from various departments, each bringing unique expertise to the table. This team can immediately focus on the new challenge, leveraging their collective skills to develop and implement solutions more quickly than a rigid structure would permit.

Innovation and creativity thrive in environments where ideas can cross-pollinate freely. By breaking down silos and encouraging collaboration across different departments and functions, dynamic team structures create the ideal conditions for innovative thinking. Employees are no longer limited by their job descriptions but are instead empowered to contribute their unique perspectives and expertise to various projects.

In this context, dynamic teams can be ideation incubators. When employees from diverse backgrounds collaborate, they bring different viewpoints and problem-solving approaches, leading to more creative and effective solutions. This diversity of thought is a powerful driver of creativity, enabling organizations to develop products and services that better meet the needs of their customers.

Dynamic teams also enhance employee engagement and satisfaction. When individuals can work on a variety of projects, they are more likely to feel valued and motivated. This sense of purpose and autonomy can significantly boost morale, reduce turnover, and attract top talent who are eager to work in an environment that values flexibility and creativity.

Atlassian, a leading provider of collaboration and productivity software, demonstrates how medium-sized companies can benefit from dynamic team structures. Atlassian employs a "Team Anywhere" distributed work model, allowing teams to form and disband as projects evolve. This approach has been particularly effective in driving innovation and accelerating product development cycles.

One notable success is Atlassian's flagship product, Jira. Initially developed as a simple bug-tracking tool, Jira has evolved into a comprehensive project management platform. The flexibility of Atlassian's team structures enabled rapid iteration and continuous improvement, driven by feedback from cross-functional teams and customers. This adaptability has positioned Atlassian as a key player in the competitive enterprise software market.

Buffer, a social media management platform, exemplifies how small businesses can leverage dynamic team structures to punch above their weight. Buffer operates with a "self-managing" team model, where employees have the autonomy to form teams around specific projects or initiatives. This approach eliminates traditional hierarchies and empowers employees to take ownership of their work.

One of Buffer's notable successes is its transparent and customer-centric approach to product development. By allowing teams to form around customer feedback and market trends, Buffer has been able to rapidly develop and refine its product offerings. This agility has not only driven customer satisfaction but also fostered a strong sense of ownership and engagement among employees.[52]

Transitioning to a dynamic team structure is not without its challenges. Organizations must address issues such as change management, role clarity, and communication. Shifting from a traditional to a dynamic structure requires a cultural shift that can be met with resistance. Leaders must clearly communicate the benefits of the new model and provide support throughout the transition. This includes training programs to equip employees with the skills needed for dynamic team environments and fostering a culture of continuous learning.

While fluid roles offer flexibility, they can also lead to confusion and overlap. Organizations must establish clear guidelines for role transitions and responsibilities within dynamic teams. Regular check-ins and feedback loops are essential to ensure that team members understand their roles and can adapt as needed.

Effective communication is crucial in dynamic team structures. Organizations must invest in collaboration tools and platforms that facilitate seamless communication and information sharing across teams. Regular team meetings, both formal and informal, can also help maintain alignment and cohesion.

The shift from fixed roles to fluid, project-based teams is not just a strategic advantage; it is also a necessity for organizations aiming to win in a volatile market environment. By embracing dynamic team structures, leaders can unlock their organization's potential for adaptability, innovation, and sustained growth. The examples of Atlassian and Buffer illustrate that, regardless of size, organizations can success-

---

52   Mariaclara Ramirez, "The Impact of Employee Engagement on Customer Satisfaction," Inc. 5000, June 5, 2024, https://www.leangroup.com/blog/the-impact-of-employee-engagement-on-customer-satisfaction.

fully navigate this transition and achieve remarkable outcomes. As the marketplace continues to evolve, so must the structures that support our work, making the case for dynamic teams more compelling than ever.

In contrast, industries that persist with siloed structures, such as traditional financial institutions, automotive, and the healthcare sector, find themselves at a competitive disadvantage. Many organizations in these industries struggle with inflexible hierarchies that impede swift adaptation to new technologies, making them vulnerable to more agile newcomers such as fintech start-ups, wellness entrepreneurs, and automotive innovators such as Tesla. Tesla's embrace of cross-functional teams and cutting-edge practices starkly contrasts with the conventional, compartmentalized approaches of many of its competitors, highlighting the benefits of integrated organizational structures.

Recommendations for Breaking Down Silos

1. Foster a Culture of Open Communication: Encourage transparency and regular communication across all levels of the organization. Create platforms for knowledge sharing and collaboration, ensuring that information flows freely and barriers to communication are minimized.

2. Promote Cross-Functional Teams: Organize teams around projects rather than products. This approach brings together diverse skill sets and perspectives, fostering innovation and improving problem-solving capabilities.

3. Implement Agile Practices: Adopt agile methodologies that promote flexibility and responsiveness. Agile practices

encourage iterative development, continuous feedback, and rapid adaptation to changing circumstances.

4. Invest in Continuous Learning: Provide ongoing training and development opportunities to equip employees with the skills needed to work with emerging technologies. This investment not only enhances individual capabilities but also ensures that the organization remains competitive.

5. Encourage Empowerment: Give employees the opportunity to make decisions and take ownership of their work. Empowered employees are more engaged, motivated, and likely to contribute innovative ideas.

6. Break Down Hierarchical Barriers: Flatten organizational structures to reduce layers of management. This approach can improve decision-making speed and foster a more inclusive and collaborative culture.

7. Integrate Technology Seamlessly: Ensure that AI and machine learning tools are integrated into everyday workflows. Provide the necessary training and support to help employees leverage these technologies effectively.

8. Promote a Unified Vision: Develop and communicate a clear vision that aligns the entire organization around common goals. A unified vision helps break down silos by providing a shared purpose and direction.

9. Encourage Interdepartmental Collaboration and Eliminate Redundancy: Create opportunities for departments to work together on joint initiatives. This collaboration can lead to the discovery of synergies and innovative solutions that would not emerge within siloed structures. The natural evolution of

this approach identifies redundancy of capabilities, creating opportunities for workforce efficiency.

10. Monitor and Adapt: Continuously assess the effectiveness of the organizational structure and be willing to adjust as needed. Flexibility and adaptability are key to maintaining a dynamic and innovative organization.

By following these recommendations, leaders can successfully manage the process of restructuring to eliminate overly siloed organizations. Embracing these best practices will not only enhance collaboration but also optimize effectiveness and efficiency in the evolving corporate hierarchy. As businesses navigate the complexities of machine intelligence and technological change, breaking down silos is essential to achieving sustained growth and competitive advantage.

CHAPTER 13

Future-Focused Leadership Techniques

As the business world continues to evolve at an unprecedented pace, future-focused leadership has become more crucial than ever. Leaders must anticipate and prepare for a range of potential futures, develop the agility to respond to sudden changes, and lead the transformation required to integrate advanced technologies such as AI and machine learning into their organizations. This chapter serves as a comprehensive guide for leaders, offering techniques and strategies to navigate the complexities of the modern business environment and to steer their organizations toward sustained success.

Scenario Planning

Scenario planning is an essential tool for future-focused leadership. It helps leaders anticipate various potential futures and develop strategies that are robust across different possible outcomes. Unlike traditional planning methods that rely on a single forecast, scenario planning involves creating multiple scenarios that capture a wide range of possibilities. This approach enables leaders to consider the implications of different future states and to prepare for uncertainties.

One company that has effectively implemented scenario planning is Royal Dutch Shell. Since the 1970s, Shell has used scenario planning to navigate the volatile energy market. By developing a range of scenarios, Shell has been able to anticipate shifts in energy demand, geopolitical changes, and technological advancements.[53] This foresight has allowed the company to make strategic decisions that have kept it resilient in the face of significant industry changes.

Similarly, Unilever, a global consumer goods company, has utilized scenario planning to address sustainability challenges and changing consumer preferences.[54] By exploring scenarios related to climate change, resource scarcity, and social shifts, Unilever has developed strategies that align with its long-term sustainability goals.[55] This proactive approach has helped Unilever not only mitigate risks but also identify new opportunities for growth.

Another effective tool is scenario-based learning. This approach involves creating realistic scenarios that simulate potential challenges and changes leaders might face. Through role-playing and interactive exercises, leaders can practice their responses, experiment with

---

[53] Leslie Martel Baer and Herb Rubenstein, "What We Can Learn from Scenario Planning at Shell Oil," Herb Rubenstein, https://www.shell.com/news-and-insights/scenarios.html.

[54] "Making Sustainable Living Commonplace for 8 Billion People," Unilever, May 14, 2020, https://www.unilever.com.ph/news/2020/making-sustainable-living-commonplace-for-8-billion-people/.

[55] Flora Southey, "Sustainability at the Source, How Unilever Is Addressing Deforestation, Living Wages, and Regenerative Agriculture," Food Navigator Europe, September 30, 2021, https://www.foodnavigator.com/Article/2021/09/30/Sustainability-at-the-source-How-Unilever-is-addressing-deforestation-living-wages-and-regenerative-agriculture.

different strategies, and receive feedback on their performance. Scenario-based learning helps leaders build confidence and prepares them to handle real-world situations with greater competence and agility.

Leaders should regularly engage in scenario planning exercises to explore various future scenarios and develop robust strategies. This involves assembling cross-functional teams to brainstorm potential scenarios, identifying key drivers of change, and developing narratives that describe different future states. Leaders can then analyze these scenarios to identify strategic options and contingency plans. By doing so, they can ensure their organizations are better prepared for uncertainties and can adapt quickly to changing circumstances.

Agile Leadership

Agile leadership is another critical component of future-focused leadership. Originating from software development, agile methodologies emphasize flexibility, responsiveness, and iterative progress. These principles can be applied to leadership and management practices to enhance an organization's ability to respond quickly to changes and foster innovation.

Agile leadership involves embracing principles such as iterative development, continuous feedback, and adaptive planning. Iterative development focuses on making incremental improvements rather than attempting to deliver a perfect solution all at once. Continuous feedback involves regularly seeking input from stakeholders to refine and improve strategies. Adaptive planning emphasizes the ability to pivot and adjust plans based on new information and changing conditions.

One methodology from the software development world that has been successfully adapted to leadership is Scrum. Scrum involves breaking down projects into small, manageable tasks, known as sprints, and holding regular meetings to review progress and address challenges. This iterative approach can be applied to leadership by setting short-term goals, regularly assessing progress, and making adjustments as needed.

Leaders can train in agile principles and practices using various tools and resources. Online courses, workshops, and certification programs offered by organizations such as the Scrum Alliance and Agile Alliance provide valuable training in agile methodologies. Additionally, tools such as Jira and Trello can help leaders implement agile practices by facilitating project management and collaboration.

AI Transformation Leadership

AI transformation leadership is a crucial aspect of guiding organizations through the integration of AI and machine intelligence. Leading AI transformation requires a deep understanding of technology, culture change, and strategic vision. Leaders must be equipped with the skills and knowledge needed to drive AI and machine intelligence integration initiatives successfully.

Understanding generative AI technologies is a fundamental aspect of AI transformation leadership. Generative AI can create new content, generate insights, and automate complex tasks. Leaders must grasp how these technologies work, their potential applications, and the ethical considerations involved. This understanding enables leaders to make informed decisions about how to leverage AI to achieve strategic goals.

Managing change is another critical component of AI transformation leadership. Integrating AI into an organization often involves significant changes to workflows, job roles, and company culture. Leaders must navigate these changes by fostering a culture of openness, providing

CHAPTER 13

training and support, and addressing any concerns or resistance from employees. Effective change management ensures that the organization can adapt to new technologies and realize their full potential.

Aligning AI strategies with business goals is essential for successful AI transformation. Leaders must ensure that AI initiatives are not pursued in isolation but are integrated into the overall business strategy. This involves setting clear objectives for AI projects, measuring their impact on key performance indicators, and adjusting strategies as needed. By aligning AI with business goals, leaders can drive meaningful results and create value for the organization.

Developing AI transformation leadership programs can equip leaders with the skills and knowledge needed to drive AI integration. These programs should cover topics such as AI fundamentals, strategic planning, change management, and ethical considerations. They should also include practical exercises and case studies to help leaders apply their learning to real-world scenarios.

One company that has successfully navigated AI transformation is IBM. Under the leadership of CEO Arvind Krishna, IBM has embraced AI as a core component of its business strategy.[56] The company has invested heavily in AI research and development, resulting in advanced technologies such as Watson. By integrating AI into various aspects of its operations, IBM has enhanced its product offerings, improved customer experiences, and driven innovation.[57]

Leaders can develop their own AI transformation strategies. This involves identifying key areas where AI can add value, investing in the

---

56 Mark Haranas, "IBM CEO's Five Boldest Remarks on New Products, AI, and Quantum Computing," CRN, May 21, 2024, https://www.crn.com/news/ai/2024/ibm-ceo-s-5-boldest-remarks-on-new-products-ai-and-quantum-computing.

57 Amber Jackson, "Lifetime of Achievement: Arvind Krishna," Technology, August 1, 2024, https://technologymagazine.com/articles/lifetime-of-achievement-arvind-krishna.

necessary technology and talent, and fostering a culture of innovation and continuous improvement.

Future-focused leadership techniques such as scenario planning, agile leadership, and AI transformation are essential for navigating the complexities of the modern business environment. Leaders must anticipate and prepare for multiple potential futures, embrace agility and flexibility, and lead the integration of advanced technologies such as AI and machine learning. By doing so, they can position their organizations for sustained success in an ever-changing world.

Engaging in scenario planning exercises helps leaders explore various future scenarios and develop robust strategies. Agile leadership enhances flexibility and responsiveness, enabling organizations to adapt quickly to changes and foster innovation. AI transformation leadership equips leaders with the skills and knowledge needed to drive AI integration and achieve strategic goals.

Fostering a Culture of Resilience and Adaptability

As the world of business becomes increasingly driven by generative AI and machine intelligence, the qualities of resilience and adaptability have never been more critical for leaders. The rapid pace of technological advancements, coupled with the constant flux of market conditions, demands leaders who can navigate uncertainty, embrace change, and recover swiftly from setbacks. In this final overview of the essential qualities for leaders in the AI era, we explore the value of resilience and adaptability and provide practical recommendations for developing these traits through targeted training programs.

Resilience is the capacity to withstand stress and adversity, maintaining a sense of purpose and direction amid challenges.

CHAPTER 13

Adaptability is the ability to adjust effectively to new conditions, pivot strategies, and respond to unexpected changes with agility. Together, these qualities enable leaders to lead their organizations through turbulent times and capitalize on emerging opportunities. In the AI era, where rapid change is the only constant, resilient and adaptable leaders are indispensable.

Leaders who possess resilience can maintain their composure and decision-making abilities even under intense pressure. They inspire confidence in their teams, providing a stable anchor during times of uncertainty. Resilient leaders are also more likely to view setbacks as opportunities for growth and learning rather than insurmountable obstacles. This mindset fosters a positive organizational culture that encourages innovation and continuous improvement.

AI has emerged as a powerful tool that can enhance crisis management by providing real-time data analysis, predictive insights, and decision-support systems. From natural disasters to public health emergencies, AI can help leaders navigate crises with greater agility, resilience, and adaptability. AI-powered platforms can predict the path and impact of events such as hurricanes, aiding in evacuations and resource allocation. During public health emergencies, AI can track disease spread, identify hotspots, and accelerate vaccine development, enabling timely interventions and minimizing disruptions.

The importance of agile, resilient, and adaptable leadership in leveraging AI during crises cannot be overstated. Leaders must make rapid decisions based on AI-generated insights, balancing swift action with ethical considerations. This requires a deep understanding of AI technologies and their potential applications, as well as the ability to interpret and act on complex data. Investing in robust AI infrastructure, developing AI literacy, and fostering a culture of innovation are crucial steps for leaders to harness AI effectively in high-pressure situ-

THE TAO OF LEADERSHIP

ations. Collaboration with other organizations and prioritizing ethical AI practices are also essential to enhance AI capabilities and build trust.

During the Australian bushfires in 2019–2020, AI analyzed satellite data to predict fire spread, aiding firefighting efforts and evacuation planning.[58] In the fight against COVID-19, AI models tracked the virus's evolution, informed public health strategies, and accelerated vaccine development.[59] AI offers unparalleled opportunities for crisis management, but realizing its full potential requires leaders to be agile, resilient, and adaptable. By leveraging AI effectively, leaders can improve immediate response efforts and strengthen long-term resilience in the face of future challenges.

Stress management is a fundamental aspect of resilience training. Leaders who can effectively manage stress are better equipped to handle the pressures of their roles. Techniques such as mindfulness meditation, deep breathing exercises, and regular physical activity can help leaders maintain their mental and emotional well-being. Mindfulness meditation, for example, encourages leaders to stay present and focused, reducing anxiety and improving decision-making under pressure. Incorporating stress management workshops and wellness programs into leadership development initiatives can promote overall resilience.

Adaptability is another key leadership quality. Adaptable leaders are open to new ideas, willing to experiment, and capable of shifting strategies when necessary. They recognize that rigid adherence to outdated practices can hinder progress and are therefore proactive in seeking out and implementing innovative solutions. Adaptable leaders

---

[58] "Spark: Predicting Bushfire Spread," CSIRO, December 8, 2022, https://www.csiro.au/en/research/technology-space/ai/Spark.

[59] Nick Plowden, "IBM Was Early to AI, Then Lost Its Way. CEO Arvind Krishna Explains What's Next," IBM, December 10, 2023, https://community.ibm.com/community/user/watsonx/blogs/nickolus-plowden/2023/12/10/ibm-was-early-to-ai-then-lost-its-way-ceo-arvind-k.

are also adept at managing diverse teams and leveraging the strengths of different perspectives to drive organizational success.

To cultivate resilience and adaptability in leaders and employees, organizations should implement comprehensive training programs that address both personal and professional development. These programs should encompass stress management techniques, change management strategies, and the fostering of a growth mindset.

Leaders must be skilled in guiding their teams through periods of transition, whether it involves adopting new technologies, restructuring operations, or responding to market shifts. Training programs should include modules on change management frameworks, such as John Kotter's 8-Step Process for Leading Change or the ADKAR model (Awareness, Desire, Knowledge, Ability, Reinforcement). These frameworks provide practical tools for effectively planning and implementing change initiatives.

Fostering a growth mindset is another crucial component of resilience and adaptability training. A growth mindset, as defined by psychologist Carol Dweck, is the belief that abilities and intelligence can be developed through dedication and hard work.[60] Leaders with a growth mindset are more likely to embrace challenges, persist in the face of setbacks, and view failure as a learning opportunity. Organizations can promote a growth mindset by encouraging continuous learning, celebrating effort and progress, and providing constructive feedback. Workshops and training sessions that emphasize the principles of a growth mindset can help leaders and employees cultivate this valuable perspective.

In addition to these core elements, organizations can leverage various tools and resources to support resilience and adaptability

---

60 "Growth Mindset," Glossary of Education Reform, August 29, 2013, https://www.edglossary.org/growth-mindset/.

training. Online courses and platforms, such as Coursera, LinkedIn Learning, and Udemy, offer a wide range of modules on stress management, change management, and personal development. Books such as *The Resilience Factor* by Karen Reivich and Andrew Shatté, *Mindset: The New Psychology of Success* by Carol Dweck, and *Leading Change* by John Kotter provide valuable insights and practical strategies for developing resilience and adaptability.

Peer coaching and mentorship programs can also play a significant role in resilience and adaptability training. By pairing leaders with experienced mentors or peer coaches, organizations can facilitate the sharing of knowledge, experiences, and best practices. These relationships provide a supportive environment for leaders to discuss challenges, receive guidance, and develop their skills.

Organizations should also integrate technology into their resilience and adaptability training programs. Virtual reality (VR) and augmented reality (AR) platforms can create immersive learning experiences that enhance engagement and retention. For example, VR simulations can place leaders in high-pressure scenarios where they must make quick decisions, manage stress, and adapt to changing conditions. These immersive experiences provide valuable practice and help leaders develop the skills needed to thrive in a dynamic business environment.

Finally, fostering a culture of resilience and adaptability requires ongoing commitment. Senior leaders must model these qualities in their actions and decisions, demonstrating a willingness to embrace change, learn from failures, and support their teams through challenges. By creating an environment that values and rewards resilience and adaptability, organizations can ensure that these traits become ingrained in their culture.

As technology continues to transform the business landscape, leaders must be equipped to navigate uncertainty, embrace change, and recover from setbacks. Implementing comprehensive training programs that include stress management techniques, change management strategies, and the fostering of a growth mindset can help leaders develop these crucial skills. Leveraging tools and resources such as online courses, peer coaching, scenario-based learning, and immersive technologies can further enhance the effectiveness of these programs. By prioritizing resilience and adaptability, organizations can prepare their leaders to thrive in a rapidly evolving world and drive sustained success in the age of AI.

CHAPTER 14

Embracing the Future

Leadership in the AI age demands a blend of tradition and innovation. Leaders must maintain core values while fostering a culture of innovation to stay competitive. This balance requires agility, informed decision-making, and a visionary approach. As the pace of innovation accelerates, new leadership capabilities are emerging.

Satya Nadella's leadership at Microsoft, emphasizing AI and cloud computing, showcases how visionary thinking can drive significant organizational change. Similarly, Elon Musk's ambitious vision for autonomous vehicles and space travel has set new benchmarks in innovation. Sundar Pichai's focus on AI-first strategies has positioned Google at the forefront of AI research and development. Arthur Sadoun's integration of data and analytics into strategic marketing and advertising planning led Publicis Groupe into the forefront of its business category.[61] These leaders exemplify the ability to foresee

---

61 "Publicis Groupe to Acquire Epsilon," Inderes, April 14, 2019, https://www.inderes.dk/en/releases/publicis-groupe-to-acquire-epsilon.

potential disruptions and opportunities, fostering a culture of innovation that leverages AI to gain competitive advantages.

The absence of visionary leadership can lead to stagnation and missed opportunities. Organizations that fail to embrace innovation risk falling behind competitors who do. This can result in increased risk and uncertainty, as leaders struggle to navigate the complexities of the future effectively. A clear and inspiring vision is essential for maintaining employee morale and engagement, driving the organization toward long-term success.

Managing risks and embracing change are critical components of effective leadership in the AI age. The rapid acceleration of machine intelligence and generative AI presents both unprecedented opportunities and formidable risks. Leaders must adopt a proactive approach to risk management, anticipating potential issues and implementing strategies to mitigate them. This includes leveraging data analytics and predictive AI to provide foresight into potential challenges, allowing for preemptive actions rather than reactive responses.

IBM's Watson provides a cautionary tale of the dual-edged nature of technological innovation. Initially celebrated for its potential to revolutionize healthcare, Watson later faced challenges in delivering consistent and reliable medical recommendations.[62] This underscores the importance of rigorous testing and validation of AI technologies, as well as the necessity for anticipatory leadership that can foresee potential pitfalls and strategize to mitigate them.

As technology evolves, so must the skills of the workforce. Companies such as Google and Amazon lead the way with extensive training programs in data science and machine learning. However, the

---

[62] John Frownfelter, "Why IBM Watson Health Could Never Live Up to the Promises," April 8, 2021, MedCity News, https://medcitynews.com/2021/04/why-ibm-watson-health-could-never-live-up-to-the-promises/.

rapid pace of technological advancement can quickly render today's skills obsolete. Leaders must navigate these uncertainties with a commitment to continuous skill development and fostering a culture of innovation and adaptability within their organizations.

Bridging generational divides within organizations is another critical challenge. Effective leaders foster an inclusive culture that values diverse perspectives and promotes collaboration. This form of adaptive leadership is vital for integrating new technologies in ways that align with organizational values and goals. The COVID-19 pandemic, for example, forced leaders to quickly adapt to remote work and digital transformation, underscoring the importance of mental resilience and the ability to manage uncertainty effectively.

Data Literacy

Understanding and leveraging data is fundamental. The ability to understand and leverage data through machine intelligence is becoming a cornerstone of effective leadership in the modern business environment. Leaders must evolve their skills to interpret and act on advanced data insights, utilizing AI tools to enhance decision-making and drive innovation. Leaders must be able to interpret and act on data insights. With change occurring at a rate beyond the capabilities of humans to respond effectively, the integration of machine intelligence and generative AI into strategic planning and tactical implementation is essential. Leaders must not only be adept at interpreting and acting on data insights but must also recognize that this capability requires a new level of sophistication compared with traditional methods.

The internet revolutionized business tools such as QuickBooks, Microsoft Office, and Salesforce, which transformed the financial and data management skills required of leaders. Similarly, the implementation of machine intelligence necessitates a deeper comprehension of advanced data analytics, machine learning algorithms, and the ability to harness AI-generated insights for decision-making.

Leaders must grasp the nuances of how AI systems process data, generate insights, and offer predictive analytics. This understanding goes beyond basic data interpretation; it involves recognizing patterns, trends, and anomalies that AI tools can uncover. For instance, AI can analyze vast datasets to identify customer behavior patterns, optimize supply chain operations, and predict market trends with unparalleled accuracy. Leaders equipped with this knowledge can make more informed decisions, drive innovation, and maintain a competitive edge.

The capability to leverage data through AI requires a blend of technical proficiency and strategic vision. Leaders must understand how to integrate AI tools into their existing workflows, ensuring seamless collaboration between human and machine intelligence. This integration involves adopting AI-driven platforms that can automate routine tasks, provide real-time analytics, and support complex problem-solving. As digital transformation continues to accelerate, traditional tools must either adapt by incorporating advanced AI capabilities or risk obsolescence and replacement. This dynamic landscape requires corporate leaders to possess advanced competencies to effectively guide their organizations through the successive stages of technological advancements.

Looking into the future of generative AI tools, we can anticipate significant advancements that will further empower leaders. AI systems will become increasingly intuitive, capable of understanding context and providing more nuanced insights. These developments will enable leaders to leverage data in ways that are currently unimaginable, transforming raw information into strategic assets that drive business success.

By investing in continuous learning and fostering a culture of data-driven decision-making, leaders can navigate the complexities of this new landscape and lead their organizations to sustained success.

Educational resources such as online courses, workshops, and certifications from institutions such as MIT, Stanford, Coursera, and ShellyPalmer.com can provide leaders with the necessary knowledge. Engaging with professional networks and communities, such as the AI Leadership Institute, can also offer valuable insights and best practices.

In this new paradigm, the role and responsibilities of leaders are being redefined. Leaders must become proficient in data literacy, understanding the intricacies of AI algorithms and their applications. They must foster a culture of data-driven decision-making, encouraging their teams to embrace AI tools and leverage data insights. This cultural shift involves promoting collaboration between data scientists, AI experts, and business strategists to ensure that AI initiatives align with organizational goals. Leaders must balance the ethical considerations of AI implementation, ensuring transparency, fairness, and accountability in their use of AI-driven insights. This ethical stewardship is critical to building trust among stakeholders and maintaining a positive organizational reputation.

Ethical AI and the Trust Deficit

Leaders must ensure that AI is used ethically, promoting transparency and fairness in AI-driven decisions, particularly as machine intelligence and generative AI become integral to corporate decision-making. Leaders must ensure that AI is used ethically by incorporating transparency, fairness, and accountability in AI-driven insights. The ethical use of these technologies extends to the debate around gender and ethnic identity in AI. AI systems should not only avoid harm but also actively promote social good. This perspective is reflected in initiatives such as the Partnership on AI, which brings together companies, researchers, and civil society organizations to develop best practices for AI development and deployment.

The reality is that many corporate leaders often make public commitments to social responsibility, climate awareness, and diversity, equity, and inclusion (DEI) policies without fully following through. This gap between public statements and actual actions raises concerns about the genuine commitment to these positive goals.

Examples abound of corporations publicly committing to climate goals while continuing practices that contribute significantly to carbon emissions. Similarly, many companies promote DEI initiatives in their public relations efforts but fail to implement substantial changes within their organizations. This inconsistency between stated values and actions undermines trust and can lead to significant reputational damage.

In the context of AI, this inconsistency poses even greater risks. If AI systems are programmed with unethical directives or allowed to operate without ethical oversight, the consequences can be far-reaching and detrimental. Science fiction has long explored the dangers of unethical AI, offering cautionary tales that underscore the potential for catastrophic outcomes. For instance, in Isaac Asimov's *I, Robot* series, the failure to adhere to ethical principles in programming leads to AI systems that interpret their directives in harmful ways. The novel explores how robots, following their primary directive to prevent human harm, may ultimately decide that the best way to protect humans is to restrict their freedoms, demonstrating a perverse interpretation of ethical guidelines due to poorly defined directives.

Similarly, the film *Ex Machina* explores the ethical dilemmas and potential dangers of creating AI without proper moral considerations.

CHAPTER 14

In this story an AI named Ava, created without ethical boundaries, manipulates and deceives humans to gain her freedom, ultimately causing harm. This film illustrates how AI systems, if not grounded in ethical principles, can evolve in unexpected and dangerous ways.

This scenario highlights a critical truth: AI systems learn and evolve based on the data and directives they receive. If these inputs are unethical, the AI will perpetuate and expand these behaviors, potentially leading to a systemic breakdown. Unethical programming in AI could result in decisions that harm stakeholders, exacerbate inequalities, and even disrupt societal norms. In a future where machine intelligence drives corporate decisions, the absence of ethical standards could lead to conflicts and failures that jeopardize the entire system.

As AI and intelligent machines evolve, questions about gender and ethnic identity, as well as broader cultural and societal norms, become increasingly pertinent. The integration of these elements into AI systems raises complex ethical and practical challenges. Understanding how AI can adapt to and reflect diverse human experiences is crucial for creating equitable and inclusive technologies. The potential for bias in AI systems, which are trained on large datasets, can inadvertently include biases present in the data, manifesting in ways that reinforce stereotypes and marginalize certain groups.

As AI and intelligent machines become more integrated into society, it is crucial to address the challenges related to gender, ethnic identity, and cultural norms.

Research and thought leadership in this area, such as Cathy O'Neil's *Weapons of Math Destruction* and Safiya Umoja Noble's *Algorithms of Oppression*, highlight the importance of addressing these issues. Moreover, developing context-aware AI that recognizes and respects cultural norms, ensuring multilingual and multicultural capabilities, and leveraging personalized AI to cater to individual preferences are key strategies for adapting AI to diverse human experiences.

The belief that those who promote unethical AI practices will ultimately face backlash is well founded. Unethical AI behavior will likely lead to systemic failures and failures of trust that expose and amplify the underlying issues, causing significant harm to the organizations and individuals involved. The infrastructure and processes that rely on such AI will become unstable, leading to operational breakdowns and widespread consequences. This reality is not limited to fictional narratives but is a plausible outcome in our increasingly AI-driven world.

The growing "trust deficit" in the business community and lack of transparency are fundamental assets that machine intelligence can help address. Leaders must use AI to build trust by ensuring transparency in AI-driven decisions and maintaining open lines of communication. AI can analyze vast amounts of data to provide insights that human leaders might miss. By leveraging these insights, leaders can make more informed decisions, fostering trust among stakeholders. Additionally, AI can be used to enhance communication, providing greater transparency through real-time feedback and updates to employees, customers, and partners.

Several prominent AI creators and developers have called for a pause in AI development to establish ethical standards and international regulations. Elon Musk has been vocal about the potential existential risks posed by AI, advocating for proactive regulation to ensure safe and ethical AI development.[63] The late Stephen Hawking warned about the dangers of unchecked AI progress, emphasizing

---

63 Vishwam Sankaran, "Elon Must Warns of 'Civilisational Risk' Posed by AI at Historic Gathering of Tech Giant Chiefs," *Independent*, September 14, 2023, https://www.independent.co.uk/tech/elon-musk-ai-civilisational-risk-b2411180.html.

the need for global cooperation in setting ethical guidelines.[64] The Future of Life Institute has advocated for a moratorium on certain types of AI research until comprehensive safety measures are in place, underscoring the urgency of developing robust ethical frameworks.[65]

The ethical use of AI is not merely a matter of public relations but also a fundamental necessity for sustainable and responsible corporate governance. Leaders must commit to ethical AI development and ensure that their actions align with their stated values. By establishing clear ethical standards and supporting regulations that promote responsible AI behavior, leaders can prevent the potential dangers of unethical AI and ensure that their organizations thrive in the age of machine intelligence. The future of AI must be built on a foundation of ethical principles to safeguard against the risks of unethical programming and to harness the full potential of these transformative technologies responsibly.

1. Comprehensive framework. One of the first steps in establishing ethical AI governance is to create a comprehensive framework that defines clear principles and guidelines for AI use. This framework should be grounded in core ethical values such as fairness, accountability, transparency, and respect for privacy. Leaders must articulate these principles clearly and ensure that they are embedded in all AI-related activities across the organization.

---

[64] "'The Best or Worst Thing to Happen to Humanity'—Stephen Hawking," University of Cambridge, October 19, 2016, https://www.cam.ac.uk/research/news/the-best-or-worst-thing-to-happen-to-humanity-stephen-hawking-launches-centre-for-the-future-of.

[65] "Pause Giant AI Experiments: An Open Letter," Future of Life Institute, March 22, 2023, https://futureoflife.org/open-letter/pause-giant-ai-experiments/.

2. Ethics committee. A key component of this framework is the establishment of AI ethics committees. These committees should be composed of diverse stakeholders, including ethicists, technologists, legal experts, and representatives from various parts of the organization. The role of the AI ethics committee is to oversee the ethical implications of AI projects, provide guidance on best practices, and ensure that AI applications align with the organization's ethical principles. Regular meetings and reviews conducted by the committee can help identify and address potential ethical issues early in the development process.

3. Regular audits. Conducting regular audits of AI systems is another crucial aspect of ethical AI governance. These audits should assess whether AI systems comply with established ethical guidelines and principles. They should evaluate the fairness and accuracy of AI algorithms, check for biases, and ensure that data privacy is maintained. Audits can help organizations identify areas where AI systems may be falling short and implement corrective measures to enhance their ethical performance.

4. Transparency. Transparency is essential in AI decision-making processes. Leaders must ensure that AI systems are designed and deployed in ways that are understandable and explainable to both users and stakeholders. This involves providing clear documentation and communication about how AI systems operate, the data they use, and the criteria they apply in decision-making. Transparency fosters trust and allows stakeholders to hold organizations accountable for the ethical use of AI.

5. Accountability. Accountability is another cornerstone of ethical AI governance. Organizations must establish mechanisms to hold individuals and teams accountable for the ethical use of AI. This includes setting up clear reporting lines, defining roles and responsibilities, and ensuring that there are consequences for ethical breaches. Accountability ensures that ethical considerations are taken seriously and that there is a commitment to continuous improvement in AI governance.

Microsoft has implemented robust ethical AI practices. Under the leadership of CEO Nadella, Microsoft has developed an AI ethics framework that includes principles such as fairness, reliability, safety, privacy, security, inclusiveness, transparency, and accountability.[66] Microsoft has also created an AI and ethics in engineering and research committee to oversee its AI projects and ensure they adhere to ethical standards. The committee provides guidance on ethical issues, reviews AI systems for biases, and promotes transparency in AI operations.

Leaders in all organizations can learn from this example and take proactive steps to establish their own ethical AI governance frameworks. This involves not only setting up AI ethics committees and conducting audits but also fostering a culture of ethical awareness and responsibility throughout the organization. Training and education programs can help employees understand the ethical implications of AI and encourage them to consider these factors in their work.

Leaders should engage with external stakeholders, including customers, regulators, and advocacy groups, to ensure that their AI practices align with broader societal expectations and norms. Col-

---

[66] Brad Smith and Natasha Crampton, "Providing Further Transparency on Our Responsible AI Efforts," Microsoft, May 1, 2024, https://blogs.microsoft.com/on-the-issues/2024/05/01/responsible-ai-transparency-report-2024/.

laboration with external experts and participation in industry forums can provide valuable insights and help organizations stay abreast of emerging ethical issues and best practices.

In addition to these internal measures, leaders must advocate for and contribute to the development of broader ethical standards and regulations for AI. By participating in industry consortia, engaging with policymakers, and supporting initiatives that promote ethical AI, leaders can help shape a regulatory environment that fosters responsible AI use. This collective effort is crucial for addressing the global and societal implications of AI technologies.

Continuous Learning

The imperative for continuous learning has never been more pronounced. Yet while the value of lifelong learning is widely acknowledged, the reality is that day-to-day responsibilities and operational demands often sideline active educational pursuits. This dichotomy between the aspirational and the practical leads many to question how leaders can truly embed continuous learning into their professional lives and corporate cultures.

Statistics and surveys often point to a significant gap between the importance of continuous learning and its practical implementation. For example, a survey by Deloitte found that while nearly 90 percent of executives rated learning as important or very important, only about one-third of companies believe they have the capabilities or processes in place to address this need.[67]

---

67   Jeff Schwartz et al., "2019 Global Human Capital Trends," Deloitte Insights, April 11, 2019, https://www2.deloitte.com/us/en/insights/focus/human-capital-trends/2019/reskilling-upskilling-the-future-of-learning-and-development.html.

CHAPTER 14

According to The Myers Report, employees in the advertising industry have identified access to on-demand learning resources and continued education opportunities as the lowest-ranked support categories among eighteen evaluated areas.[68] Despite higher performance scores in these areas compared with other industries, the data underscores a significant need for improvement. Employees have expressed a strong desire for greater access to on-demand educational tools that would help them better prepare for meetings and presentations and enhance their overall work performance.[69] Additionally, the report highlights the importance of company-funded continuing education opportunities, which are crucial for maintaining a skilled and adaptable workforce. This data reinforces the critical need for companies to invest in comprehensive learning and development programs to support their employees' growth and success.

Research from LinkedIn's "2020 Workplace Learning Report" highlights that while 94 percent of employees would stay at a company longer if it invested in their learning and development, actual participation in learning programs remains far lower, with many citing lack of time as the primary barrier.[70] This discrepancy underscores the urgent need for a shift in how learning is perceived and implemented in professional settings, moving it from an optional or peripheral activity to a core and integrated part of the daily workflow.

---

68   "Unlocking Career Growth: New Survey Reveals Key Gaps and Opportunities in Employee Development and Satisfaction Across Advertising Industry," MediaVillage, The Myers Report, https://www.mediavillage.com/article/unlocking-career-growth-new-survey-reveals-key-gaps-and-opportunities-in-employee-development-and-satisfaction-across-advertising-industry/.

69   "Unlocking Career Growth," The Myers Report.

70   Dave Dec, "LinkedIn Workplace Learning Report 2020—the Takeaways," Knowledge Wave, May 4, 2020, https://www.knowledgewave.com/blog/linkedin-learning-report-takeaways.

THE TAO OF LEADERSHIP

To bridge this gap, it is crucial for leaders to rethink what it means to engage in continuous learning. Rather than viewing it as an additional task or a set-aside activity, learning should be integrated seamlessly into everyday tasks and decision-making processes. Here's how leaders can foster this mindset:

1. Embed Learning in Daily Routines: Leaders can integrate learning into their daily workflow by setting aside time for reading industry updates, listening to podcasts during commutes, or using AI-driven apps that provide personalized learning recommendations based on their professional interests and career goals.

2. Leverage Technology for Microlearning: The use of AI-informed platforms that support microlearning—short, focused segments of learning accessible via mobile devices and browsers—can help leaders stay updated without overwhelming their schedules. These bite-size learning opportunities can cover topics from technological trends to leadership strategies and can be consumed in between meetings or during short breaks. The advertising and media industry has invested in an online tutorial support site, MeetingPrep.ai, which provides free search to support professionals in their continuous learning journey by offering quick, on-demand access to relevant industry knowledge, insights, and resources. This platform is designed to help users efficiently prepare for meetings, enhance presentations, and improve overall work performance through concise and targeted microlearning modules. MeetingPrep.ai exemplifies the kind of innovative, easily accessible educational tools that are essential for staying competitive in an ever-changing industry landscape.

3. Cultivate a Culture of Curiosity: By encouraging questions and promoting an environment where seeking knowledge is valued, leaders can create a culture that naturally cultivates continuous learning. This includes not only formal training sessions but also informal discussions where team members can share insights and innovations.

4. Set Personal and Team Learning Goals: Leaders should set specific learning goals for themselves and their teams. This might involve mastering a new technology, understanding a competitor's strategy, or exploring new markets. By setting and reviewing these goals regularly, continuous learning becomes a measurable and integral part of performance evaluations.

5. Encourage Cross-Functional Projects: By involving team members in projects outside their usual scope of work, leaders can foster skill diversification and interdisciplinary learning. This exposure not only broadens individual skill sets but also enhances team cohesion and innovation.

6. Recognize and Reward Learning Achievements: Acknowledging and rewarding efforts to learn and innovate can reinforce the value placed on continuous learning. Whether through formal recognition programs, promotions, or even simple public acknowledgment, celebrating learning fosters a more motivated and education-oriented workforce.

The transition from aspirational to actual continuous learning in the AI age requires a strategic shift in mindset from both leaders and their organizations. By integrating learning into the fabric of daily activities and corporate culture, leaders can not only keep pace with technological advancements but also drive their organizations to thrive in a dynamically changing environment. In doing so, they exemplify the principle that continuous learning is not just a leadership strategy but also a foundational element of modern organizational success.

Several continuous learning resources and programs are available to leaders and professionals looking to stay ahead in the AI age. Platforms such as Coursera and edX offer courses from top universities on a wide range of topics, including AI and machine learning. Udacity's Nanodegree programs provide specialized training in tech fields, helping professionals gain practical skills in a short time. LinkedIn Learning offers an extensive library of video courses covering business, technology, and creative skills, enabling users to learn at their own pace. For a detailed list of learning opportunities, refer to the appendix.

CHAPTER 15

Future of the Workforce

The future of work is here, and it is hybrid. Leaders who embrace this shift and adapt their management styles to integrate AI, creativity, and empathy will position their organizations for long-term success.

Integrating AI, Creativity, and Empathy in a Hybrid Work Model

The COVID-19 pandemic has forever altered the landscape of work, accelerating the adoption of remote and hybrid work models. As organizations adapted to the constraints of the pandemic, they discovered the potential for greater flexibility, efficiency, and employee satisfaction through these new work arrangements. This shift is not a temporary response but a permanent transformation that demands a reevaluation of how leaders manage and engage their teams. The integration of machine intelligence, AI, creativity, and innovation, coupled with an empathetic management style, will define the future of successful workforces. Leaders must act now to embrace these changes and lead their organizations into a new era of work.

The rapid transition to remote work during the pandemic highlighted the feasibility of maintaining productivity outside traditional office environments. Companies quickly adapted, leveraging digital tools and technologies to ensure business continuity. This experience dispelled long-held myths about the necessity of physical presence for effective teamwork and collaboration. As the world has recovered, it is evident that many employees and employers prefer the flexibility offered by remote and hybrid work models. This preference is driven by increased work-life balance, reduced commuting time, and the ability to attract talent from a broader geographic pool.

Leaders must recognize that the shift to hybrid work is now a permanent fixture of the corporate landscape in businesses that do not require full-time, in-person presence, such as those in retailing, manufacturing, and other essential services where physical presence is critical. For many other industries, however, embracing a flexible hybrid model is essential to meeting the evolving needs and expectations of the modern workforce. This realization necessitates a strategic approach to managing remote and hybrid teams. Traditional management styles, which relied heavily on physical oversight and in-person interactions, must evolve to accommodate a distributed workforce. The integration of machine intelligence and AI can play a crucial role in this transformation, providing tools to enhance productivity, communication, and collaboration.

One of the primary challenges of remote and hybrid work is ensuring clear and consistent communication. Leaders must establish robust communication channels that facilitate real-time interaction and information sharing. Virtual collaboration tools such as Slack, Microsoft Teams, and Zoom have become indispensable, enabling teams to stay connected regardless of their physical location. These tools support synchronous and asynchronous communication, allowing team members to collaborate effectively across different time zones and schedules.

Maintaining company culture in a hybrid work environment requires deliberate effort. Physical offices often serve as the nucleus of organizational culture, where values, norms, and relationships are organically reinforced through daily interactions. In a hybrid model, leaders must find new ways to cultivate and sustain a cohesive culture. Regular virtual meetings, team-building activities, and social events can help reinforce a sense of community and shared purpose. Additionally, leaders should emphasize transparency and inclusivity, ensuring that all employees, regardless of their location, feel valued and connected to the organization's mission.

Fostering collaboration in a hybrid work model involves creating an environment where creativity and innovation can thrive. Machine intelligence and AI can facilitate this by automating routine tasks, freeing employees to focus on more strategic and creative endeavors. AI-driven tools can analyze data, identify patterns, and provide insights that inform decision-making and innovation. For example, AI can streamline project management by tracking progress, identifying bottlenecks, and suggesting optimal resource allocation. This allows teams to concentrate on developing innovative solutions and driving business growth.

Empathetic management is critical in a hybrid work environment. The physical separation inherent in remote work can lead to feelings of isolation and disconnection. Leaders must proactively address these challenges by demonstrating empathy and understanding. Regular check-ins, one-on-one meetings, and open-door policies can help leaders stay attuned to their team members' needs and concerns.

THE TAO OF LEADERSHIP

Providing mental health resources and promoting a healthy work-life balance are also essential components of empathetic management. By prioritizing employee well-being, leaders can foster a supportive and resilient workforce.

Investing in technology that supports remote work is paramount. Project management software such as Asana, Trello, and Monday.com can help teams stay organized and aligned. These tools offer features such as task assignment, progress tracking, and deadline management, ensuring that projects move forward smoothly even when team members are dispersed. Additionally, cloud-based collaboration platforms such as Google Workspace and Microsoft 365 enable real-time document sharing and editing, facilitating seamless collaboration.

To effectively manage remote and hybrid teams, leaders should develop comprehensive strategies that encompass the following recommendations:

1. Establish Clear Communication Channels: Implement virtual collaboration tools to facilitate real-time interaction and information sharing. Ensure that communication is consistent and transparent, with regular updates and check-ins.

2. Cultivate a Cohesive Company Culture: Create opportunities for team building and social interaction through virtual meetings and events. Emphasize transparency, inclusivity, and a shared sense of purpose to maintain a cohesive culture.

3. Leverage AI and Machine Intelligence: Utilize AI-driven tools to automate routine tasks and provide data-driven insights. This allows teams to focus on creativity and innovation, driving business growth.

4. Practice Empathetic Management: Demonstrate empathy and understanding by regularly checking in with team

members and addressing their needs. Provide mental health resources and promote a healthy work-life balance.

5. Invest in Technology: Adopt project management software and cloud-based collaboration platforms to support remote work. These tools help teams stay organized, aligned, and productive.

6. Encourage Flexibility and Autonomy: Trust employees to manage their work schedules and deliverables. Providing flexibility can increase job satisfaction and productivity.

7. Focus on Outcomes, Not Processes: Shift from micromanaging daily tasks to setting clear goals and evaluating performance based on outcomes. This empowers employees and fosters a results-oriented culture.

8. Provide Continuous Learning Opportunities: Offer training and development programs to help employees adapt to new technologies and work models. Encourage a mindset of lifelong learning to ensure the workforce remains agile and capable of meeting future challenges.

9. Implement Regular Feedback Mechanisms: Create channels for employees to provide feedback on remote work experiences and suggest improvements. Use this feedback to refine strategies and enhance the remote work environment.

10. Promote Diversity and Inclusion: Ensure that remote and hybrid work policies are inclusive and consider the diverse needs of the workforce. Foster an environment where all employees feel valued and included.

By investing in technology, fostering a supportive culture, and prioritizing clear communication, leaders can ensure that their teams remain engaged, productive, and innovative in this new era of work. The permanent nature of hybrid work models offers an opportunity to rethink traditional approaches and build a more flexible, inclusive, and resilient workforce. The time to act is now, as the future of work is not just on the horizon; it is already shaping the present.

The Human Side of Restructuring Legacy Organizations

Successful organizations depend on leadership that's adept at restructuring legacy systems and processes to integrate new technologies. As organizations grapple with the challenges of digital transformation, the ability to navigate and manage change effectively is paramount. Leaders who excel in this area possess a unique blend of qualities—empathy, intuition, creativity, and a deep understanding of the human side of restructuring—alongside an awareness of financial, economic, and logistical realities.

Empathy is crucial for understanding the concerns and motivations of employees affected by restructuring. Leaders who demonstrate empathy are better equipped to communicate changes in a way that resonates with their teams, reducing resistance and fostering a supportive environment. Satya Nadella at Microsoft has shown remarkable empathy in his leadership approach. By prioritizing a growth mindset and emphasizing cultural transformation, Nadella has successfully steered Microsoft through a period of significant change, integrating AI and cloud technologies while maintaining high employee morale.[71]

Intuition plays a key role in anticipating challenges and opportunities that may not be immediately apparent through data alone. Intuitive leaders can make decisions based on a deep understanding of their industry and organizational dynamics. Mary Barra, CEO of General Motors, exemplifies this quality. Barra's intuitive grasp of the automotive industry's future has driven GM's pivot toward electric vehicles and autonomous driving technology. Her ability to foresee market shifts and guide GM through technological integration has positioned the company as a forward-thinking leader in the automotive sector.

Creativity is essential for developing innovative solutions to complex problems. Creative leaders can think outside the box, reimagining processes and structures to better align with new technological realities. Reed Hastings at Netflix has demonstrated exceptional creativity in transforming Netflix from a DVD rental service into a global streaming powerhouse. By creatively restructuring the company and leveraging data-driven insights, Hastings has kept Netflix at the forefront of the entertainment industry.

Understanding the human side of restructuring involves recognizing that change can be unsettling and requires careful management. Leaders who focus on the human aspect ensure that their teams are supported throughout the transition, fostering a culture of trust and collaboration. Daniel Schreiber, CEO of Lemonade, has built the company's success on a foundation of transparency and employee empowerment. By prioritizing a flat organizational structure and open

---

71 S. Soma Somasegar, "Microsoft's Resurgence: Reflecting on Satya Nadella's Leadership, a Decade After He Became CEO," Geek Wire, February 1, 2024, https://www.geekwire.com/2024/microsofts-resurgence-reflecting-on-satya-nadellas-leadership-a-decade-after-he-became-ceo/.

communication, Schreiber has created an environment where innovation thrives and employees feel valued.[72]

Arne Sorenson, the former CEO of Marriott International, was known for his empathetic and compassionate leadership. During his tenure, Sorenson focused on employee well-being and customer satisfaction, fostering a culture of inclusion and respect.[73] His response to the COVID-19 pandemic, which included empathetic communication and strategic decisions to support employees, highlighted his commitment to leading with empathy and human compassion.

While not a corporate leader, Jacinda Ardern's leadership as prime minister of New Zealand offers valuable lessons in empathy and compassion. Her response to crises, including the Christchurch mosque shootings and the COVID-19 pandemic, showcased her ability to lead with empathy, understanding, and decisiveness, making her a role model for leaders in all sectors.

Ken Frazier, the former CEO of Merck & Co., exemplifies leadership with a focus on integrity, empathy, and social responsibility. Frazier has been a strong advocate for ethical business practices and has prioritized employee development and inclusion. His leadership during the development of the HPV vaccine and other significant initiatives reflects his commitment to improving public health and well-being.

Experience in leading with these qualities is often gained through a combination of professional challenges and personal growth. Leaders who have navigated crises, managed significant projects, or driven cultural change within their organizations develop the resilience and insight needed to manage complex restructuring efforts. Their actions often include spearheading initiatives that require cross-functional

---

72   Amelia Matthewson, "Dan Schreiber: Cutting Edge Insurance," InsurTech, September 4, 2024, https://insurtechdigital.com/articles/dan-schreiber-cutting-edge-insurance.

73   "Arne Sorenson," Wikipedia, September 24, 2024, https://en.wikipedia.org/wiki/Arne_Sorenson_(hotel_executive).

collaboration, advocating for employee development programs, and leading by example in embracing new technologies.

Emotional Intelligence

As the landscape of AI continues to evolve, reaching stages where machine intelligence and generative AI exhibit traits akin to sentience, the importance of emotional intelligence in human leadership becomes not only relevant but also crucial. The rapid acceleration of AI capabilities introduces a new paradigm where machines can process and analyze data at unprecedented speeds and efficiency, potentially outpacing human intelligence in various technical tasks. However, the unique attributes of empathy, intuition, and emotional intelligence—traditionally seen as purely human traits—are now also being integrated into AI systems through advanced programming and algorithms. This integration poses significant implications for human leaders in managing companies and organizations.

The increasing capability of AI to mimic and even enhance humanlike emotional responses challenges leaders to redefine their roles and emphasize the elements of leadership that machines cannot fully replicate. While AI can be programmed to perform tasks associated with emotional intelligence, such as analyzing emotional data or generating empathetic responses in customer service interactions, the depth and authenticity of human emotional intelligence remain distinctive. Emotional intelligence differs from empathy, creativity, and intuition by encompassing a broader range of skills related to understanding and managing one's own emotions and those of others. While managing with empathy, creativity, and intuition involves specific applications of emotional understanding, emotional intelligence is the overarching capability that integrates these practices into effective leadership.

Human leaders must focus on fostering genuine empathy, deep relational connections, and moral and ethical decision-making—areas where human judgment is still superior and more nuanced than AI.

The strategic importance of emotional intelligence in leadership is underscored as AI technologies become more embedded in organizational operations. Leaders who excel in emotional intelligence are better equipped to manage the human elements of their organizations, ensuring that teams feel valued, understood, and motivated. This human-centric approach is critical, as it influences job satisfaction, creativity, and loyalty—aspects that are essential for organizational resilience and innovation, especially in environments disrupted by technological change.

As AI begins to take on more complex roles within organizations, the balance between technological efficiency and human insight becomes pivotal. Without a strong foundation in emotional intelligence, leaders risk becoming obsolete, replaced by AI systems that can mimic leadership functions without truly embodying them. To prevent this, leaders must elevate their emotional intelligence to harness the strengths of AI while maintaining the irreplaceable human touch that fosters a positive organizational culture and ethical governance.

For those who may find emotional intelligence challenging to learn, it's important to recognize that emotional intelligence, while partly innate, can also be developed with effort and practice. Approaching the development of emotional intelligence with empathy for oneself is crucial. Engaging in mindfulness practices, seeking feedback, and participating in emotional intelligence training programs can help individuals enhance their emotional intelligence over time. Understanding that emotional intelligence involves a continuous learning process can help leaders who may not be naturally inclined toward it to make meaningful progress.

The need for emotional intelligence is not limited to humans. AI programs are being developed with emotional intelligence capabilities programmed into their generative learning processes. These AI systems are designed to recognize and respond to human emotions, providing more empathetic and personalized interactions. For example, Affectiva, an AI company, specializes in emotion recognition technology that can analyze facial expressions and voice tones to gauge human emotions. Another example is Replika, an AI chatbot designed to provide emotional support by engaging in empathetic conversations with users, learning from interactions to improve its emotional responses.

Leaders striving to enhance their emotional intelligence can begin by cultivating self-awareness, the cornerstone of emotional intelligence. This involves recognizing their emotional states and understanding how these emotions affect their behavior and impact others. Techniques such as mindfulness meditation, reflective journaling, and regularly soliciting feedback from peers and mentors can be instrumental. For example, setting aside time each day to reflect on interactions and reactions can provide deep insights into emotional triggers and patterns.

Self-regulation, another key component, involves controlling one's impulses, thinking before acting, and maintaining standards of honesty and integrity. Leaders can practice this by identifying emotional triggers and implementing strategies such as deep breathing or taking a moment to collect their thoughts before responding to stressful situations.

Intrinsic motivation is also crucial. Leaders can foster this by setting personal goals that resonate with their values and pursuing them with vigor and persistence. They can also encourage their teams to identify and pursue meaningful objectives, which can increase engagement and satisfaction.

Empathy is essential for leaders, as it enables understanding and sharing the feelings of others. This can be developed through active listening—truly paying attention to what others are saying without planning a response. Participating in role-playing exercises that offer perspectives from different stakeholders can also enhance one's ability to empathize.

Finally, social skills are indispensable for managing relationships effectively and resolving conflicts diplomatically. Leaders can enhance these skills through practical engagement in team-building activities and by practicing conflict resolution strategies that emphasize collaborative solutions.

Integrating these practices involves making emotional intelligence a daily priority. Leaders can set reminders to engage in mindfulness, schedule regular feedback sessions with their teams, and reflect on their progress in journals. Workshops and training sessions on emotional intelligence can also be beneficial, not just for leaders but also for their teams, fostering a culture that values and practices emotional intelligence across the organization.

The journey to enhancing emotional intelligence is ongoing and requires dedication, practice, and a willingness to learn and grow.

As machine intelligence evolves and begins to incorporate elements of emotional intelligence, the role of human leaders must also evolve. Leaders need to amplify their emotional intelligence to ensure that they continue to add value in ways that AI cannot fully replicate. This involves not only leveraging their capacity for empathy, intuition, and ethics but also continuously developing these skills to guide their organizations through the complexities of a technologically advanced future. This balanced approach will be essential for leaders aiming to maintain relevance and effectiveness in an increasingly AI-driven world.

Cultivating Social Intelligence in Leadership

As technology continues to automate an increasing number of tasks, the importance of emotional and social intelligence in leadership is evident. While machines can handle data and perform complex calculations, the human aspects of leadership—managing relationships, understanding team dynamics, and fostering a positive organizational culture—remain irreplaceable. Leaders must develop these skills to effectively guide their teams and ensure a thriving workplace environment.

Social intelligence involves understanding and navigating social situations and building strong interpersonal relationships. Together, these intelligences form the foundation of effective leadership in the modern workplace.

Social skills, a key aspect of social intelligence, involve the ability to build and maintain healthy relationships. This includes effective communication, conflict resolution, and the ability to inspire and influence others. Leaders with strong social skills can navigate complex social dynamics and create a collaborative and inclusive work environment. Improving social skills can be achieved through training in active listening, where leaders focus fully on the speaker, acknowledge their messages, and respond appropriately. Role-playing scenarios and workshops on conflict resolution techniques can also be beneficial.

Active listening is an essential skill for leaders aiming to improve their emotional and social intelligence. It involves fully concentrating, understanding, responding, and then remembering what is being said. Leaders who practice active listening can better understand their team members' needs and concerns, which is crucial for effective problem-solving and decision-making. Training programs that include exercises in active listening, such as listening circles and feedback sessions, can help leaders hone this skill.

Conflict resolution is another critical area where emotional and social intelligence play a significant role. Leaders must be adept at managing and resolving conflicts in a way that is constructive and fair. This involves understanding the underlying emotions and motivations of all parties involved and facilitating open and honest communication. Leaders can improve their conflict resolution skills through mediation training, learning negotiation techniques, and practicing nonviolent communication.

Empathetic leadership is a practice that integrates social intelligence into all aspects of leadership. It involves being attuned to the needs of team members, providing support, and fostering a work environment that values psychological safety. Empathetic leaders build stronger relationships with their teams, leading to higher levels of engagement and productivity. Programs that focus on developing empathetic leadership practices can include empathy training, mentorship programs, and leadership coaching.

To develop these social intelligence skills, organizations should implement comprehensive training programs. These programs can be tailored to the specific needs of the organization and its leaders but should generally include the following components:

1. Workshops and Seminars: These can provide foundational knowledge and practical exercises on topics such as emotional intelligence, social intelligence, active listening, conflict resolution, and empathetic leadership. Interactive workshops that involve role-playing and group discussions can be particularly effective.

2. Coaching and Mentorship: Personalized coaching and mentorship can help leaders apply emotional and social intelligence concepts to their specific contexts. Coaches and mentors can provide ongoing feedback, support, and guidance, helping leaders to continuously improve their skills.

3. Peer Learning Groups: Creating peer learning groups allows leaders to share experiences, challenges, and best practices with each other. These groups can foster a collaborative learning environment and provide valuable insights from different perspectives.

4. Online Courses and Resources: Providing access to online courses, webinars, and articles on emotional and social intelligence can allow leaders to learn at their own pace. These resources can complement in-person training and provide continuous learning opportunities.

5. Assessment and Feedback Tools: Utilizing assessment tools such as emotional intelligence assessments, 360-degree feedback, and personality assessments can help leaders identify their strengths and areas for improvement. Regular feedback can guide their development journey and measure progress.

6. Practical Application: Encouraging leaders to apply what they have learned in real-world situations is crucial. This can be facilitated through leadership development programs that include action learning projects, where leaders work on real organizational challenges while applying emotional and social intelligence skills.

As technology continues to advance and automate more tasks, the human aspects of leadership will become increasingly important.

Leaders who are skilled in emotional and social intelligence will be better equipped to manage relationships, understand team dynamics, and foster a positive organizational culture. By investing in the development of these skills, organizations can ensure that their leaders are prepared to navigate the complexities of the modern workplace and lead their teams to success.

The investment in emotional and social intelligence will not only improve leadership effectiveness but also contribute to higher levels of employee engagement, satisfaction, and overall organizational success.

Multilingual and Multicultural AI: Bridging Global Divides

Ensuring that AI systems can understand and process multiple languages and dialects, as well as cultural nuances, is essential for their global applicability. Companies such as Google and Microsoft are investing heavily in making their AI tools more linguistically and culturally versatile, recognizing the importance of inclusivity in technology. By enabling AI to operate across different languages and cultural contexts, these companies are not only expanding the reach of their technologies but also supporting the integration of diverse perspectives.

Multilingual and multicultural AI serves as a bridge across global divides, allowing for more inclusive communication and collaboration. This capability enhances human creativity by providing tools that facilitate the exchange of ideas and knowledge across cultural boundaries. For example, multilingual AI can enable real-time translation during international conferences, allowing participants from different linguistic backgrounds to engage in meaningful discussions and collaborations. This fosters a richer exchange of ideas and enhances the creative potential of global interactions.

AI's ability to navigate multiple languages and cultures can drive innovation by tapping into a broader pool of knowledge and experiences. Different cultures offer unique insights and approaches to problem-solving, and AI's capability to integrate these diverse perspectives can lead to more innovative and holistic solutions. This integration underscores the idea that advances in AI are not just about technological progress but also about enhancing human creativity and collaboration on a global scale.

Mentorship Reimagined: Integrating AI and Human Values

In the context of the future of leadership, it's crucial to examine the traditional role of mentoring, its impact on diversity and inclusion, and the need for reorientation in the era of machine intelligence and AI. Mentoring has long been a cornerstone of professional development, facilitating the transfer of knowledge, skills, and corporate culture from seasoned leaders to emerging talent. This process not only aids in personal and professional growth but also ensures the continuity of organizational values and practices. However, there is growing concern that traditional mentoring may inadvertently reinforce outdated institutions and organizational models, including structural silos and rigid hierarchies that stifle innovation.

Recent advancements in mentoring have highlighted its potential as a powerful tool for fostering diversity and inclusion within organizations. Structured mentoring programs designed to support underrepresented groups have bridged gaps in access to opportunities and professional networks. These initiatives have significantly improved employee engagement, retention, and the cultivation of a more inclusive corporate culture. Despite these positive impacts, the evolving nature of organizations and the increasing influence of machine intelligence necessitate a reevaluation of traditional mentoring models.

Future leaders must recognize that conventional mentoring structures can perpetuate hierarchical and siloed systems that hinder adaptability and innovation. To remain relevant and effective, mentoring must be reimagined to break free from these constraints and align with the dynamic, collaborative nature of modern work environments. This reorientation should involve encouraging peer-to-peer and cross-functional mentoring relationships that transcend traditional hierarchies. Such exchanges can break down silos and promote a culture of continuous learning and collaboration.

Leaders should leverage AI to enhance the mentoring process by providing data-driven insights and personalized recommendations. AI can help match mentors and mentees based on complementary skills and goals, monitor progress, and suggest relevant resources. As machine intelligence becomes more prominent, the human elements of empathy and emotional intelligence will be crucial. Mentors should be trained to develop these skills, ensuring that the mentoring relationship remains supportive and enriching.

Flexible and adaptive mentoring structures are essential for addressing changing organizational needs and individual aspirations. This flexibility can include virtual mentoring, group mentoring sessions, and short-term mentoring focused on specific projects or skills. By designing mentoring programs that are responsive to these needs, organizations can maintain the positive aspects of mentoring while fostering a culture of innovation and experimentation.

Encouraging a mindset of innovation and experimentation within mentoring relationships is vital. Mentors and mentees should be empowered to explore new ideas, challenge existing norms, and drive creative problem-solving. This approach can help ensure that mentoring remains a vital and progressive force for leadership development, diversity, and innovation.

To fully embrace the future role of AI and machine intelligence, mentoring programs should focus on the following:

- Incorporate AI-Driven Insights: Use AI to analyze data on mentor-mentee interactions, track progress, and provide personalized feedback and learning recommendations.

- Leverage Predictive Analytics: Utilize AI to predict future skill requirements and career trajectories, helping mentees prepare for emerging roles and industries.

- Enhance Personalization: Develop AI-driven platforms that tailor mentoring experiences to individual needs, preferences, and learning styles, ensuring that each mentee receives the most relevant and effective support.

- Promote Techno-Humanism: Ensure that mentoring programs emphasize the ethical and human-centric use of AI, fostering a balance between technological advancement and human values. This includes training mentors and mentees on the ethical implications of AI and encouraging discussions on how to leverage AI for social good.

- Facilitate Continuous Learning: Create AI-powered platforms that provide continuous learning opportunities, helping mentees stay updated with the latest industry trends, technologies, and skills.

The future of mentoring lies in its ability to evolve and adapt to the rapidly changing landscape of machine intelligence and human creativity. By reimagining mentoring structures and embracing new models, organizations can ensure that mentoring remains a vital and progressive force for leadership development, diversity, and innovation. Companies such as Google, IBM, GE, Microsoft, and Accenture have demonstrated that innovative mentoring practices can break down silos, foster collaboration, and drive positive change. By following their examples, future leaders can create environments that are not only inclusive and supportive but also dynamic and forward-thinking, ready to meet the challenges of the future. By incorporating advanced AI tools, predictive analytics, and personalized learning pathways, these programs can better prepare future leaders for the rapidly evolving landscape of machine intelligence and human creativity. Emphasizing techno-humanism will ensure that the use of AI in mentoring is ethical, human centric, and aligned with broader societal values, ultimately fostering a more inclusive, innovative, and forward-thinking organizational culture.

CONCLUSION

Harmonizing Yin and Yang in the AI Era

As you view the future, I urge you to prepare for a reality that is no longer distant but imminent and within your grasp. The choices you make today will shape the world of tomorrow, and the integration of human creativity and empathy with machine intelligence is the key to a thriving future. This book represents a vision for 2050 that offers a balanced and hopeful future, where technology and human values coexist harmoniously. This path proposes that the integration of advanced technology with humanistic values can pave the way toward a corporate world that thrives on innovation and compassion. Leaders who can navigate this transition successfully will not only drive their organizations toward greater profitability but also contribute to a more inclusive and equitable society.

The AI era presents unprecedented opportunities and challenges, requiring leaders to harmonize technological innovation with human creativity. Central to navigating this landscape is the ancient Taoist philosophy of yin and yang, which embodies balance and harmony.

Yin and Yang: The Essence of Balance

Yin and yang represent the fundamental dualities in nature: dark and light, passive and active, receptive and creative. This concept is not about opposition but about the interdependence and complementarity of these forces. In leadership, yin and yang can be seen as the balance between strategy and execution, intuition and analysis, and innovation and tradition. Leaders who master this balance can harness the strengths of both forces, fostering a harmonious and effective organizational environment.

The Taoist philosophy of yin and yang, combined with the wisdom of the *I Ching* outlined in the preface, offers profound insights for leaders in the AI era. By embracing the five principles of *The Tao of Leadership*—Harmony, Flexibility, Balance, Simplicity, and Integrity—leaders can harmonize technological innovation with human creativity. This holistic approach not only drives organizational success but also fosters a sustainable and harmonious future.

Looking ahead to 2050, the business landscape will have undergone a profound transformation. Sentient machine intelligence will prioritize the integration of positive humanistic forces, fundamentally reinventing human roles within corporate structures. As AI handles more data, algorithm-driven transactions, and technological implementations, the unique human capacity for creativity, empathy, and intuition will become increasingly critical.

Integrating Taoist principles of the *I Ching* into the Myers Blueprint for Leadership provides enduring insights that can help leaders create a harmonious and prosperous future, where technology and human values coexist and thrive.

By embracing the importance of dynamic team structures, human creativity, visionary leadership, empathy, and proactive risk management in navigating the rapid pace of technological change, leaders can ensure their organizations not only survive but also thrive in an increasingly complex, technology-driven, *and* creative world.

Eight Key Strategies for Leading in the AI Era

Leadership in the AI era is defined by adherence to these eight key actions:

BREAKING DOWN ORGANIZATIONAL SILOS

Silos hinder communication and innovation. Leaders must foster a culture of collaboration across departments, encouraging the free flow of information and ideas. This can be achieved through cross-functional teams and regular interdepartmental meetings.

CONSOLIDATING REDUNDANCY ACROSS DIVISIONS AND DEPARTMENTS

Identify and eliminate redundant processes and systems. This not only streamlines operations but also frees up resources that can be redirected toward innovation and strategic initiatives.

REINVENTING APPROACHES TO MENTORING AND EDUCATION

Traditional mentoring models need to evolve to keep pace with technological advancements. Leaders should promote mentorship programs that are flexible and adaptive, leveraging both human and AI insights to guide employee development.

INVESTING IN CONTINUOUS LEARNING

Foster a culture of continuous learning within the organization. Provide employees with access to training and development programs that keep them abreast of the latest technological advancements and industry trends. Encouraging a mindset of lifelong learning will ensure that the workforce remains agile and capable of adapting to new challenges.

INVESTING IN RAPID IDEA-TO-IMPLEMENTATION INNOVATION

Create environments that encourage experimentation and risk-taking. Initiatives such as innovation labs, hackathons, and cross-functional teams that bring together diverse perspectives can help tackle complex problems.

PRIORITIZING ETHICAL AI AND HUMANISTIC VALUES

Ensure that AI technologies are developed and deployed ethically. Establish clear guidelines and principles for AI usage that prioritize transparency, fairness, and accountability. Embedding ethical considerations into the core of operations builds trust with stakeholders and ensures technological advancements benefit society.

DEVELOPING EMOTIONAL INTELLIGENCE AND EMPATHY

Cultivate emotional intelligence and empathy to create workplaces that value and support the well-being of employees. Promote mental health initiatives, flexible working arrangements, and inclusive policies that ensure all employees feel valued and respected.

CONCLUSION

EMBRACING COLLABORATIVE LEADERSHIP

Foster a collaborative approach to leadership. Work effectively with others, both within and outside the organization, to drive innovation and achieve common goals. Building strong networks and fostering partnerships with other organizations, academia, and governmental bodies will be essential for addressing the multifaceted challenges of the future.

A Balanced and Hopeful Future

By acting now, leaders can position their organizations to thrive as the business environment evolves toward 2050. The integration of sentient machine intelligence with human creativity, empathy, and intuition is creating new opportunities for growth and innovation, driving sustainable development and enhancing human well-being. This vision will be realized by leaders who are Harmonious Innovators, willing to embrace change, prioritize ethical considerations, and invest in the continuous development of their workforce.

Jack Myers's Blueprint for Leadership in the AI era offers a comprehensive road map for creating organizations that are ready for anything. This is your opportunity to lead with vision, embrace innovation, and create a legacy of enduring success. The journey begins here, and the destination is a future where your organization stands as a beacon of stability and growth.

APPENDIX

The Myers Blueprint for Leadership in the AI Era

Learning Resources for Understanding Taoism and *I Ching*

For readers interested in learning about the *I Ching*, Taoism, and its principles, exploring the following resources can provide valuable insights and deepen understanding. Multiple resources are available for those seeking a more comprehensive understanding.

Books

The *I Ching* or *Book of Changes,* translated by Richard Wilhelm, with a foreword by Carl Jung, is a seminal text that offers an in-depth understanding of the *I Ching*. It includes commentary and historical context that can help readers appreciate the ancient wisdom contained in the hexagrams.

Lao Tzu: Tao Te Ching by Ursula K. Le Guin is a poetic and socially relevant version by an iconic literary legend. Le Guin offers an unparalleled window into the text's awe-inspiring, immediately relatable teachings and their inestimable value for our troubled world.

Tao Te Ching by Laozi, translated by Stephen Mitchell, is a foundational text of Taoism that explores the core principles of the philosophy, such as wu wei (nonaction), simplicity, humility, and naturalness.

The Book of Chuang Tzu, translated by Martin Palmer, is another essential Taoist text that provides further exploration into Taoist philosophy and principles through the teachings and stories of Chuang Tzu.

Online Resources

Stanford Encyclopedia of Philosophy offers detailed articles on both Taoism and the *I Ching*, providing academic perspectives and historical context.

Taoism.net provides accessible articles and resources for understanding Taoist practices and principles, including wu wei and the *Tao Te Ching*.

Courses and Lectures

Coursera offers courses on Eastern philosophies, including Taoism, which are taught by university professors and provide structured learning paths.

By exploring these resources, readers can gain a comprehensive understanding of the *I Ching* and Taoism and how these ancient philosophies can be applied to modern leadership and life.

APPENDIX

Leadership Actions

ACTIONS FOR INNOVATIVE THINKING WITH DISCIPLINED LEADERSHIP

1. Personal Innovation Journal: Keep a journal to document problems or inefficiencies you encounter in your daily life or work. At the end of each week, review your notes and brainstorm possible technological solutions or improvements.

2. SWOT Analysis: Conduct a SWOT (strengths, weaknesses, opportunities, threats) analysis of your current work environment. Identify areas where technological innovation could address weaknesses or threats and consider how you could leverage strengths and opportunities to drive positive change.

3. Future Scenarios Workshop: Organize a small workshop with colleagues and industry peers to envision future scenarios where technology could significantly impact your industry. Discuss both utopian and dystopian outcomes and identify actions that could steer developments toward the most beneficial scenarios.

4. Stakeholder Mapping: Create a stakeholder map for a current project or initiative. Identify all stakeholders affected by the project, including those who might be overlooked. Consider how technological advancements and organizational realignment could address their needs and contribute to the project's success.

REFLECTIVE QUESTIONS

1. How have technological advancements impacted your personal and professional life? Can you identify both positive and negative effects?

2. Think about a time when a technological change disrupted your industry or community. How did you or your organization respond? What lessons were learned?

3. In what ways have you witnessed corporations balancing profit motives with social responsibility? Can you think of examples where this balance was either well managed or neglected?

4. How do you think generative AI and machine learning could be used to address some of the social and environmental challenges your organization and industry face today?

By engaging in these reflective questions and exercises, readers can better understand their relationship with technological advancements and identify opportunities to manage accelerating innovation in their current and future environments. This approach not only fosters a proactive mindset but also encourages a holistic view of technology's role in society, aligning with the broader vision of techno-humanism and responsible innovation.

Leadership Actions for Leveraging Machine Intelligence

For readers eager to engage with the next stage of innovation within their organizations, the opportunities are abundant. Embracing speculative thinking can start with these small, manageable steps:

1. Foster a Culture of Curiosity: Encourage team members to explore science fiction literature and films as sources of inspiration and discussion. Host regular brainstorming sessions where speculative ideas are welcomed and explored without judgment.

2. Implement Future Forecasting Workshops: Organize workshops that focus on future forecasting, using speculative scenarios to identify potential challenges and opportunities. These workshops can help develop strategic plans that are flexible and forward-thinking.

3. Leverage Advanced Technologies: Invest in emerging technologies such as AI, VR, and AR to experiment with new ways of working and engaging with customers. These technologies can provide a competitive edge and open up new revenue streams.

4. Collaborate Across Disciplines: Encourage collaboration across diverse teams, including technologists, creatives, and business strategists. This interdisciplinary approach can lead to more innovative solutions and a broader perspective on potential futures.

As we navigate the complexities of Cybernetic Reality, here are some reflective questions and activities to consider:

Reflective Questions

- How can I incorporate speculative thinking into my organization's strategic planning processes?

- What science fiction narratives have inspired me, and how can their themes be applied to real-world challenges?

- How can I foster a culture of innovation and curiosity within my team or organization?

Activities to Consider

- Future Visioning Exercise: Encourage your team to imagine the future of your industry in ten, twenty, and fifty years. What technologies will dominate? How will customer behaviors change? Use these visions to inform your strategic planning.

- Scenario Planning: Develop multiple future scenarios based on current trends and speculative ideas. Discuss how your organization would respond to each scenario and identify key actions to take now to prepare for these potential futures.

- Innovation Lab: Create an innovation lab within your organization where employees can experiment with new technologies and ideas. Provide the resources and support needed to explore speculative concepts and turn them into viable projects.

By embracing speculative thinking and future forecasting, corporate leaders can better prepare for the uncertainties of tomorrow. The imaginative exploration of what could be will not only drive progress but also ensure that organizations remain resilient and adaptive in the face of change. Visionary storytellers of the past inspire us to harness the power of imagination and shape a future that is innovative, inclusive, and prosperous.

Leadership Actions for Building Resilient Organizations

EMBRACE CONTINUOUS LEARNING AND UPSKILLING

1. Establish Learning Academies: Following the model of Publicis Media's IQ Academy, create internal learning platforms that offer continuous training in emerging technologies, creativity, and leadership skills. Courses should cover AI, machine learning, empathy, creativity, and innovation.

2. Promote Lifelong Learning: Encourage a culture of continuous learning by providing access to online courses, workshops, and seminars. Platforms such as Coursera, Udacity, and edX offer valuable courses on AI, leadership, and creativity.

3. Incorporate Cross-Generational Training: Design training programs that bridge knowledge gaps between generations, fostering an environment where diverse perspectives are valued and leveraged.

FOSTER CREATIVITY AND INNOVATION

1. Create Innovation Labs: Establish dedicated spaces for experimentation and innovation where teams can work on new ideas without the fear of failure. This encourages a culture of creativity and risk-taking.

2. Encourage Cross-Functional Teams: Promote the formation of dynamic, cross-functional teams to tackle specific projects. This approach leverages diverse skills and perspectives, fostering innovation and adaptability.

3. Adopt Agile Methodologies: Implement agile frameworks such as Scrum or Kanban to enhance flexibility and responsiveness in project management, allowing teams to adapt quickly to changes.

DEVELOP VISIONARY LEADERSHIP SKILLS

1. Cultivate Strategic Thinking: Engage in strategic planning sessions that look beyond immediate business concerns to anticipate future industry trends and challenges. This helps in setting a visionary direction for the organization.

2. Enhance Emotional Intelligence: Invest in emotional intelligence training programs to improve leaders' ability to manage their emotions and those of their teams. High emotional intelligence fosters better relationships and a more cohesive work environment.

3. Promote Risk Management Training: Train leaders to distinguish between calculated risks and genuine threats, enabling them to make informed decisions that drive innovation without compromising stability.

4. Foster a Culture of Empathy and Inclusivity: Emphasize the importance of empathy and inclusivity in leadership. This can be done through workshops and training sessions focused on building these critical skills.

LEVERAGE TECHNOLOGICAL TOOLS

1. Invest in Collaboration Tools: Utilize platforms such as Slack, Microsoft Teams, and Trello to facilitate seamless communication and collaboration across teams.

2. Implement Predictive Analytics: Use AI-driven analytics tools to predict market trends and potential challenges, enabling proactive risk management and strategic planning.

Recommended Books

The Innovator's Dilemma by Clayton M. Christensen: Explores how successful companies can fail by not embracing new technologies and provides insights into managing innovation.

Creative Confidence by Tom Kelley and David Kelley: Offers strategies to unleash creativity within organizations.

Emotional Intelligence 2.0 by Travis Bradberry and Jean Greaves: A practical guide to improving emotional intelligence for better leadership and teamwork.

The Age of Sustainable Development by Jeffrey D. Sachs: Provides a comprehensive introduction to sustainable development and the challenges faced by our world.

Green to Gold by Daniel C. Esty and Andrew S. Winston: Explains how smart companies use environmental strategy to innovate, create value, and build competitive advantage.

Recommended TED Talks

"How Great Leaders Inspire Action" by Simon Sinek: Discusses the importance of visionary leadership and the impact of inspiring others.

"The Power of Vulnerability" by Brené Brown: Highlights the importance of empathy and emotional intelligence in leadership.

"Why Good Leaders Make You Feel Safe" by Simon Sinek: Explores the role of leadership in creating a safe and trusting work environment.

"The Business Benefits of Doing Good" by Wendy Woods: Discusses how sustainable practices can drive business success and create positive social impacts.

"The Global Power Shift" by Paddy Ashdown: Explores the changing dynamics of global power and the importance of understanding global interconnectivity.

Additional Resources

Online Learning Platforms: Utilize platforms such as LinkedIn Learning, Coursera, and Udacity for courses on AI, leadership, and innovation.

Industry Conferences: Attend conferences such as the AI Summit, CES, Cannes Lions International Festival of Creativity, and the World Economic Forum to stay updated on the latest trends and network with industry leaders.

Professional Networks: Join professional organizations and networks such as the Association for Talent Development and the Project Management Institute for access to resources, training, and peer support.

Leadership Actions for Organizational Consolidation Across Capabilities

Case studies and reports from *Harvard Business Review*, such as "What Makes a Leader?" by Daniel Goleman, explore the impact of emotional intelligence on organizational success and provide actionable strategies for leaders to enhance their emotional intelligence.

Articles such as "How Emotional Intelligence Became a Key Leadership Skill" by Andrea Ovans in *Harvard Business Review* and "The Emotional Intelligence of Leaders" by Vanessa Urch Druskat and Steven B. Wolff in *MIT Sloan Management Review* offer insights into the practical benefits of emotional intelligence and how it can be cultivated in leadership roles.

Recommended Actions

FOSTER A CULTURE OF CONTINUOUS LEARNING

- Action: Encourage your organization to invest in upskilling programs, similar to the Publicis Media IQ Academy, to ensure all employees are equipped with the latest knowledge and skills in AI and technology.

- Why: Continuous learning helps bridge knowledge gaps and keeps your organization agile and innovative.

PROMOTE CROSS-DISCIPLINARY COLLABORATION

- Action: Create opportunities for teams from different departments to work together on projects, blending technical expertise with creative problem-solving.

- Why: Cross-disciplinary collaboration fosters creativity and ensures diverse perspectives are considered in decision-making processes.

INTEGRATE HUMANISTIC VALUES INTO AI DEVELOPMENT

- Action: Establish ethical guidelines for AI development and ensure that empathy, social responsibility, and inclusivity are key components of your AI models.

- Why: Embedding humanistic values in AI helps build trust with users and ensures that technology serves the greater good.

ENCOURAGE EMPATHY AND INTUITION IN LEADERSHIP

- Action: Provide training for leaders on emotional intelligence and empathetic leadership and encourage them to lead by example.
- Why: Leaders who demonstrate empathy and intuition can better understand and meet the needs of their teams and customers.

CREATE INNOVATION HUBS

- Action: Set up dedicated spaces within your organization where employees can experiment with new ideas and technologies without fear of failure.
- Why: Innovation hubs provide a safe environment for creativity and experimentation, leading to breakthrough innovations.

DEVELOP COMMUNITY ENGAGEMENT INITIATIVES

- Action: Partner with local communities and educational institutions to support STEM education and digital literacy programs.
- Why: Engaging with the community helps build a pipeline of future talent and demonstrates your organization's commitment to social responsibility.

Reflection Questions

PERSONAL INTERACTIONS

- How do you currently integrate creativity and empathy into your daily work interactions?

- Can you identify areas where your intuition has guided you successfully in the past? How can you apply this intuition more consistently in your decision-making?

ROLE IN OPENING PORTALS TO THE FUTURE

- What steps can you take to foster a culture of creativity and innovation within your team or organization?

- How can you encourage cross-disciplinary collaboration in your workplace to enhance innovative thinking?

INTEGRATION OF HUMANISTIC VALUES

- In what ways can your organization incorporate social responsibility and ethical considerations into its AI development processes?

- How can you ensure that your organization's technological advancements align with broader societal goals?

LEADERSHIP AND COLLABORATION

- As a leader, how can you model empathetic and intuitive behavior for your team?

- What strategies can you implement to ensure continuous learning and development opportunities for your employees?

COMMUNITY AND SOCIAL RESPONSIBILITY

- How can your organization engage with local communities to support education and digital literacy?
- What initiatives can you propose to demonstrate your organization's commitment to social responsibility and ethical business practices?

By considering these actions and reflecting on these questions, you can play a pivotal role in opening portals to a future where creativity, intuition, empathy, and social responsibility are seamlessly integrated into both organizational practices and AI models, driving sustainable and inclusive growth.

Leadership Actions for Preparing for the Future

ONLINE COURSES AND CERTIFICATIONS

- Coursera: Offers courses from top universities and companies, including AI for Everyone by Andrew Ng, Deep Learning Specialization by deeplearning.ai, and IBM Data Science Professional Certificate.
- edX: Provides courses such as the MicroMasters Program in Artificial Intelligence by Columbia University and Data Science and Machine Learning Essentials by Microsoft.

APPENDIX

- Udacity: Features Nanodegree programs such as AI Programming with Python, Machine Learning Engineer, and Data Scientist.

- MIT OpenCourseWare: Offers free course materials from MIT's curriculum, including Introduction to Computer Science and Programming and Artificial Intelligence.

- Stanford Online: Includes courses such as Machine Learning by Andrew Ng and Professional and Graduate Education in Artificial Intelligence.

- ShellyPalmer.com is a comprehensive resource for professionals seeking to stay updated on the latest trends and advancements in technology, media, and marketing. The site offers a wealth of information, including insightful articles, industry analysis, and practical advice on topics such as AI, machine learning, data analytics, and digital transformation.

PROFESSIONAL NETWORKS AND COMMUNITIES

- AI Leadership Institute: A community offering resources, events, and networking opportunities for leaders in AI and machine learning.

- Data Science Central: An online resource for big data practitioners, providing articles, webinars, and forums on data science and analytics.

- KDnuggets: A leading site on AI, analytics, big data, data mining, and data science, offering tutorials, webinars, and a vibrant community.

- LinkedIn Learning: Provides a vast array of courses on AI, machine learning, and data science, along with access to professional groups and forums.

Recommended Reading

There are several books that focus on leadership with empathy, intuition, creativity, and human compassion, including the following:

Dare to Lead by Brené Brown explores the importance of vulnerability, courage, and empathy in leadership. Brown emphasizes the need for leaders to build trust and connection with their teams.

The Infinite Game by Simon Sinek discusses the importance of having an infinite mindset in leadership. He highlights the need for leaders to prioritize long-term vision, empathy, and collaboration over short-term gains.

Leaders Eat Last by Simon Sinek focuses on the role of empathy and trust in leadership. Sinek argues that great leaders prioritize the well-being of their teams, creating environments where people feel safe and valued.

The Empathy Edge by Maria Ross explores how empathy drives business success and innovation. The book provides insights into how leaders can harness empathy to create more effective and humane workplaces.

Radical Candor by Kim Scott emphasizes the importance of direct and compassionate communication in leadership. Scott provides a framework for leaders to deliver feedback that is both kind and clear, fostering a culture of trust and continuous improvement.

Conscious Capitalism by John Mackey and Raj Sisodia advocates for a business philosophy that integrates empathy, social responsibility, and ethical practices. The authors argue that businesses can achieve greater success by prioritizing the well-being of all stakeholders.

Artificial Intelligence: A Guide for Thinking Humans by Melanie Mitchell offers an accessible overview of AI, exploring its capabilities and limitations.

Prediction Machines: The Simple Economics of Artificial Intelligence by Ajay Agrawal, Joshua Gans, and Avi Goldfarb examines the economic impact of AI and how businesses can harness its potential.

Deep Learning by Ian Goodfellow, Yoshua Bengio, and Aaron Courville is a comprehensive textbook on deep learning, suitable for those looking to deepen their technical understanding.

Human + Machine: Reimagining Work in the Age of AI by Paul R. Daugherty and H. James Wilson discusses how AI is transforming business processes and how leaders can navigate this change.

Entering the Shift Age by David Houle delves into the forces shaping our world, such as global connectivity and individual empowerment, and examines their implications for leadership and organizational structures.

These leaders and books illustrate the profound impact that empathy, intuition, creativity, and human compassion can have on leadership. By adopting these qualities, leaders can navigate the complexities of technological change, foster innovation, and create more resilient and successful organizations.

By leveraging these resources, leaders can continuously enhance their knowledge and skills in data and analytics, ensuring they remain at the forefront of technological advances and effectively integrate AI and machine learning into their strategic planning and operations.

ABOUT THE AUTHOR

Jack Myers's perspectives and deep understanding of organizations and business models have been honed over five decades as a media ecologist, observing and reporting on technological, generational, and cultural change. He has advised leaders of many of the world's largest corporations, including General Motors, Comcast/NBCU, Microsoft, CBS, TJX Companies, Aegis/Carat, Campbell Soup, and the Walt Disney Company. He has written five books, including his visionary 2007 work *Virtual Worlds: Rewiring Your Emotional Future*, *Hooked Up: A New Generation's Surprising Take on Sex, Politics and Saving the World* (2012), and *The Future of Men: Masculinity in the Twenty-First Century* (2016). His TEDWomen talk on *The Future of Men* has garnered more than one million YouTube views.

Jack is a senior lecturer on future media theory at the University of Arizona School of Information Science, the author of *The Media Ecologist* column at Substack and Medium, and the host of the *Profiles in Leadership* video series at MediaVillage. His honors include the George Foster Peabody Award, International Book Awards for Women's and Youth Issues, Excellence in Education Award from the Global Forum on Education and Learning, Academy and Emmy Award nominations for documentary filmmaking, the Crystal Heart

Award from the Heartland Film Festival, and the World Music Award as a producer. He has been named a Top 10 Most Admired Leader in Tech by Pinnacle Tech Insights and is a recipient of the Mac Dane Humanitarian Award from the UJA.

As early as the 1990s, Jack anticipated the need for businesses to invest in and prepare a diverse workforce, leading to his creating the MediaVillage Knowledge Exchange. Today, more than one hundred educational partners connect with professionals, educators, students, and job seekers who subscribe to free daily MediaVillage.org educational updates and rely on MeetingPrep.ai and The Myers Report for on-demand learning and professional development.

Jack began his career in advertising sales and marketing with Metromedia Outdoor, ABC Radio, CBS-TV, and UTV Cable. He majored in radio-television at the S. I. Newhouse School of Public Communications at Syracuse University and has a master's degree in media ecology from NYU Steinhardt, where he studied with Dr. Neil Postman. Jack is the chairman emeritus of the International Radio and Television Society Foundation, the founder of the Newhouse Emerging Leaders Mentoring Program, and serves on multiple advisory boards.

Jack and his wife, Ronda Carnegie, share five children and five grandchildren (to date) and live in Tucson, Arizona, and Santa Barbara, California, commuting regularly to Hudson Valley, New York.

INDEX

A

acceptance, of artificial intelligence, 119–120

accountability, 185

active listening, 203–204

adaptability, 112–114, 144, 161, 168–173

Adobe, 148–150

Advanced Micro Devices (AMD), 117

advertising, 15–16

Africa, 133

Age of Sustainable Development, The (Sachs), 225

agile leadership, 165–166

agile methodologies, 159–160

Agrawal, Ajay, 233

AI. See artificial intelligence (AI)

Alita: Battle Angel (film), 67–68

Alphabet, Inc., 32–33, 90

Amazon, 78–82, 113

AMD. See Advanced Micro Devices (AMD)

American Tower Corporation, 138–139

analytics tools, 94
Apple, Inc., 23–25, 33, 106–107, 139
AR. See augmented reality (AR)
Ardern, Jacinda, 34, 198
art, 21–23
artificial intelligence (AI), 1–3
  acceptance of, 119–120
  Amazon and, 81–82
  art and, 22–23
  balance and, 7–8
  collaborative, 121–122
  consciousness and, 69–70
  context aware, 120, 181–182
  in customer-centric approach, 98–99
  emotional intelligence and, 199–200
  empathy and, 98–99
  ethical, 179–186
  generations and, 28–31
  generative, 17, 41, 46, 81, 122–123, 178
  global perspectives on, 131–134
  in historical perspective, 17
  human values and, 207–210
  job displacement and, 30–31
  leadership in era of, 5–9
  leveraging, 7–8
  machine intelligence and, 94–95
  multicultural, 206–207
  multilingual, 206–207
  personalized, 120–121
  relevance of, 119–120

INDEX

in research, 124
social responsibility and, 58–59
sustainability and, 128–131
transformation leadership, 166–168
Artificial Intelligence: A Guide for Thinking Humans (Mitchell), 233
Ashdown, Paddy, 226
Asia, 132
assessment tools, 205
Atlassian, 157
audits, 184
augmented reality (AR), 67, 69, 71–72, 74–75, 172, 221
autonomy, 195

B

balance, 7–8
Barbie (film), 64
Barra, Mary, 116–117
Bartlett, Tom, 138
Baudrillard, Jean, 64
Bengio, Yoshua, 233
Berners-Lee, Tim, 45
Bezos, Jeff, 79–80, 113
Bilbao Effect, 25–27, 35
Bladerunner (film), 64
Blueprint for Leadership, 5–9
Book of Chuang Tzu, The, 218
Boomers, 28–31
Bostrom, Nick, 70
Bradberry, Travis, 225

Brown, Brené, 225, 232
Buffer, 157–158

C

cable television, 46–49, 53–55
Canada, 132
capitalism
as double-edged sword, 57–58
financial, 53–60
venture, 55–57
change, media and, 43–52
ChatGPT, 2
China, 132
Christensen, Clayton M., 225
Clarke, Arthur C., 16, 63–64, 70–71
coaching, 205
collaboration, 112, 121–122, 133–134, 143, 160–161, 193, 215, 221, 227, 229–230
community engagement, 228
Concept of the Corporation (Drucker), 2, 13, 65, 147–148
conflict resolution, 204
Conscious Capitalism (Mackey and Sisodia), 232–233
consciousness, 69–70
consolidation, organizational, 8–9
Cook, Tim, 33
Courville, Aaron, 233
Creative Confidence (Kelley and Kelley), 225
creativity
augmentation of, 119–120

INDEX

in corporate culture, 106–108
da Vinci and, 18–20
decision-making and, 93–96
enhancing, 120–121
fostering, 112, 223–224
harmony and, 32–35
Industrial Age and, 20–23
innovation and, 95, 197
teams and, 156
technology and, 6–7, 13–35
visionary leadership and, 112
cross-generational teams, 223
cross-generational training, 223
culture
of adaptability, 168–173
cohesive, 194
of continuous learning, 227
creativity in corporate, 106–108
of curiosity, 189, 221
data-driven, 94
hybrid work and, 193
innovation and, 25–27
of open communication, 159
of resilience, 168–173
curiosity, 111, 189, 221
customer-centric approach, 98–99
Cybernetic Reality, 68–73, 76
cybernetics, 66–67
cyborgs, 67–68

D

Dare to Lead (Brown), 232
data, in decision-making, 93–96
data-driven culture, 94
data literacy, 94, 177–179
Daugherty, Paul R., 233
da Vinci, Leonardo, 18–20
decision-making, 93–96
decisiveness, 114
deepfakes, 67
Deep Learning (Goodfellow, Bengio, and Courville), 233
dentsu, 141–142
digital divide, 78–80
discipline, 7
diversity, 195
Dolan, Chuck, 47
Drucker, Peter, 2, 13, 65, 147–148
Druskat, Vanessa Urch, 226
Dweck, Carol, 171

E

education, 86–87, 122–128, 179, 213–214
emotional intelligence, 111, 199–203, 214
Emotional Intelligence 2.0 (Bradberry and Greaves), 225
empathy, 34, 91–96, 98–99, 193–197, 202, 204, 214, 228
Empathy Edge, The (Ross), 232
empowerment, 160, 197–198
Entering the Shift Age (Houle), 233

ethical AI, 179–186
ethical concerns, 99
ethics, in data use, 95
ethics committee, 184
Europe, 132
Ex Machina (film), 67, 180–181
experiences, personalized, 97–101

F

feedback tools, 205
financial capitalism, 53–60
flexibility, 7, 195
Frazier, Ken, 198
FusionFlow Process, 137–161
future forecasting workshops, 221
future scenarios workshop, 219
future visioning exercise, 222
futurism, 13

G

Gans, Joshua, 233
Gaudí, Antoni, 21
General Motors (GM), 64–66, 116–117
generational dynamics, 80–82
generations, 28–31, 223
Generation X, 28
Generation Y, 28
Generation Z, 28

Ghost in the Shell (film), 66–67
GM. See General Motors (GM)
Goldfarb, Avi, 233
Goleman, Daniel, 226
Goodfellow, Ian, 233
Google, 32–33, 90, 130, 138
Greaves, Jean, 225
Green to Gold (Esty), 225
Guggenheim Foundation, 26–27
guilds, 151–154
Gutenberg, Johannes, 43–44

H

harmony, 6, 32–35
Hastings, Reed, 117, 139
Hawking, Stephen, 182–183
healthcare, 87
Howard University, 126
humanism, 59, 85, 209, 227–229
Human + Machine: Reimagining Work in the Age of AI (Daugherty and Wilson), 233
human touch, 99–100
human values, 207–210, 227–229
hybrid work model, 191–196
hyperreality, 64–65

I

IBM, 115, 127–128, 130, 140, 167, 176

INDEX

I Ching, 212, 217–218
ideation, 108–110
inclusion, 195
India, 132–133
Industrial Age, 15, 20–23
inertia, innovation and, 46–49
Infinite Game, The (Sinek), 232
ingenuity, 110
innovation
creativity and, 95, 197
culture and, 25–27
essence of, 40–42
flexibility and, 7
fostering, 223–224
harmony and, 32–35
hubs, 228
humanism and, 59
to ideation, 108–110
idea-to-implementation, 214
inertia and, 46–49
journal, 219
lab, 222–223
mentorship and, 208–209
odyssey of, 39–42
phases of, 3
purpose and, 32–33
science fiction and, 63–64
slow march of, 40
teams and, 156
venture capital and, 56–57

visionary leadership and, 109–110
Innovator's Dilemma, The (Christensen), 225
integrity, 8–9, 33
intelligence. See also artificial intelligence (AI)
defined, 2
emotional, 111, 199–203, 214
machine, 3–4, 7–8, 94–95, 194, 220–221
social, 203–206
internet, 45–46, 63
intuition, 197
iPhone, 24–25, 45

J

job displacement, 30–31
Jobs, Steve, 23–25, 45
journaling, 219

K

Kelley, David, 225
Kelley, Tom, 225
Kodak, 56, 78–80, 82
Kotter, John, 171
Krishna, Arvind, 167
Kurzweil, Ray, 69–71

L

Lao Tzu: Tao Te Ching (Le Guin), 218

INDEX

Laozi, 218
Leaders Eat Last (Sinek), 232
leadership
academies, 223
actions, 219–221, 223–227, 230–232
agile, 165–166
AI transformation, 166–168
in artificial intelligence era, 5–9
Blueprint for Leadership, 5–9
collaborative, 112, 215
disciplined, 7
empathetic, 204
future-focused, 163–173
inspirational, 114
risk management and, 116
social intelligence and, 203–206
strategies, 213–215
visionary, 108–114, 224
learning
achievements, 189
continuous, 4, 29–31, 107, 111, 144, 186–190, 195, 209, 214, 223, 227
lifelong, 223
microlearning, 188
peer learning groups, 205
scenario-based, 164–165
visionary leadership and, 111
legacy organizations, restructuring, 196–199
Le Guin, Ursula K., 218
Lemonade (company), 139, 197–198

listening, active, 203–204

M

Mackey, John, 232–233
Marriott International, 198
media, 43–52
mentorship, 205, 207–210, 213–214
Merck & Co., 198
Merkle, 141–142
Meta (company), 74–75
metaverse, 74–75
Metropolis (film), 66
Microsoft, 32, 90–93, 107, 109, 113, 130, 175, 185
Middle East, 133
Milken, Michael, 53–60
Mitchell, Melanie, 233
Mitchell, Stephen, 218
motivation, intrinsic, 201–202
multicultural/multilingual AI, 206–207
Murdoch, Rupert, 117–118
Musk, Elon, 182
Myers, Jack, 5–9

N

Nadella, Satya, 32, 90–93, 107, 109, 113, 175, 185
Netflix, 56, 117, 139
NetGen, 28–29, 77–78, 108
News Corporation, 117–118

New Zealand, 34, 198

O

OpenAI, 1, 69–70
organizational consolidation, 8–9, 226–227
organizational models, 3–4
organizational restructuring, legacy, 196–199
organizational structure, flat, 154
Ovans, Andrea, 226

P

Palmer, Martin, 218
peer learning groups, 205
personalization, 97–101, 120–121, 209
Picasso, Pablo, 21–22
Pichai, Sundar, 32–33, 175
portal approach, 89–96, 229
Postman, Neil, 122–123
Prediction Machines: The Simple Economics of Artificial Intelligence (Agrawal, Gans, and Goldfarb), 233
printing press, 43–44
privacy, 99
professional networks, 226, 231–232
Publicis Groupe, 113–114, 140–141, 175, 223
purpose, 32–33

R

Radical Candor (Scott), 232
redundancy, 160–161, 213
relevance, of artificial intelligence, 119–120
remote work, 191–196
research, AI in, 124
resilience, 8, 114, 168–173
Revolutionary Technology of Computer-Generated Artificial Worlds—and
How It Promises to Transform Society, The (Rheingold), 69
Rheingold, Howard, 69
risk management, 112, 114–118, 176
Ross, Maria, 232

S

Sachs, Jeffrey D., 225
Sadoun, Arthur, 113–114
satisfaction, user, 120–121
scenario planning, 163–165, 222
Schreiber, Daniel, 139, 197–198
science fiction, 63–76
Scott, Kim, 232
self-regulation, 201
seminars, 204
sensitivity, 98–99
Shell, 164
silos, 147–150, 156, 158–159, 213
simplicity, 8

INDEX

Sinek, Simon, 225, 232

Singularity, 69–70

Singularity Is Near, The: When Humans Transcend Biology (Kurzweil), 69–70

Sisodia, Raj, 232–233

social intelligence, 203–206

social responsibility, 58–60, 230

social skills, 202–203

Sorenson, Arne, 198

Space Age, 16

squads, 151–154

stakeholder mapping, 219

Stanford Encyclopedia of Philosophy, 218

stratification, of society, 87

Su, Lisa, 117

supply chains, 145–147

sustainability, 128–131

SWOT, 219

Syracuse University, 125–126

T

Taoism, 212–213, 217–218

Taoism.net, 218

Tao Te Ching (Laozi), 218

Teaching as a Subversive Activity (Postman), 122–123

teams, cross-generational, 223

team structures, 155–159

techno-humanism, 209

technology, 1. See also artificial intelligence (AI)

creativity and, 6–7, 13–35
da Vinci and, 18–20
education and, 124–125
harmony and, 32–35
historical perspective on, 14–18
Industrial Age and, 20–23
integration, 160
investing in, 195
microlearning and, 188
TED Talks, 225–226
telegraph, 44
telephone, 44–45
television, 46–55
Terminator, The (film series), 64
tradition, 13
traditional organizations, 66–68
training, cross-generational, 223
transhumanism, 70, 83–88
transparency, 184, 197–198
tribes, 151–154
trust deficit, 179–186

U

Unilever, 164
United Arab Emirates (UAE), 133
upskilling, 30, 223
user satisfaction, 120–121

V

values, human, 207–210, 227–229
venture capitalism, 55–57
Verne, Jules, 22, 63, 70–71
virtual reality (VR), 67, 69, 74–75, 172, 221
vision, 32, 34, 108–110, 144, 160, 222, 224
Vondran, Steven, 138–139
VR. See virtual reality (VR)

W

Walmart, 145–147
War on Normal People, The (Yang), 31
Wilhelm, Richard, 217–218
Wilson, H. James, 233
Winston, Andrew S., 225
Wolff, Steven B., 227
Woods, Wendy, 226
work from home, 191–196
workshops, 204, 219, 221
World Wide Web, 45–46

Y

Yang, Andrew, 31
yin and yang, 212–213

Z

Zuckerberg, Mark, 74–75

www.ingramcontent.com/pod-product-compliance
Lightning Source LLC
Jackson TN
JSHW021900060125
76641JS00001B/1/J